Praise for *DOPAMINE NATION*

"Anna Lembke deeply understands an experience I hear about often in the therapy room at the nexus between our modern addictions and our primal brains. Her stories of guiding people to find a healthy balance between pleasure and pain have the power to transform your life."

—Lori Gottlieb, "Dear Therapist" columnist at *The Atlantic* and
New York Times bestselling author of *Maybe You Should Talk to Someone*

"Just when you thought you knew all you needed to know about the addiction crisis, along comes Dr. Anna Lembke with her second brilliant book on the topic—this one not about a drug but about the most powerful chemical of all: the dopamine that rules the pain and pleasure centers of our minds. In an era of overconsumption and instant gratification, *Dopamine Nation* explains the personal and societal price of being ruled by the next fix—and how to manage it. No matter what you might find yourself over-indulging in— from the Internet to food to work to sex—you'll find this book riveting, scary, cogent, and cleverly argued. Lembke weaves patient stories with research, in a voice that's as empathetic as it is clear-eyed."

—Beth Macy, author of the *Washington Post* Best Book of the Year,
the *New York Times* Notable Book of 2018, and bestseller *Dopesick:
Dealers, Doctors, and the Drug Company that Addicted America*

"We all desire a break from our routines and those parts of life that upset us. What if, instead of trying to escape these things, we learn to turn toward them, to reach a peaceful harmony with ourselves and the people we share our lives with? Lembke has written a book that radically changes the way we think about mental illness, pleasure, pain, reward, and stress. Turn toward it. You'll be happy you did."

—Daniel J. Levitin, *New York Times* bestselling author of *The Organized Mind* and *Successful Aging*

"[An] eye-opening survey on pleasure-seeking and addiction . . . Readers looking for balance will return to Lembke's informative and fascinating guidance."   —*Publishers Weekly* (starred review)

"Fascinating case histories and a sensible formula for treatment."   —*Kirkus Reviews*

DOPAMINE NATION

Finding Balance in the
Age of Indulgence

ANNA LEMBKE, MD

DUTTON

DUTTON

An imprint of Penguin Random House LLC

penguinrandomhouse.com

Previously published as a Dutton hardcover in August 2021
First Dutton trade paperback printing: January 2023

THE LIBRARY OF CONGRESS HAS CATALOGED
THE HARDCOVER EDITION OF THIS BOOK AS FOLLOWS:
Names: Lembke, Anna, 1967– author.
Title: Dopamine nation : finding balance in the age of indulgence /
Anna Lembke, M.D.
Description: New York : Dutton, [2021] |
Includes bibliographical references and index.
Identifiers: LCCN 2020041077 (print) | LCCN 2020041078 (ebook) |
ISBN 9781524746728 (hardcover) | ISBN 9781524746735 (ebook)
Subjects: LCSH: Pleasure. | Pain. | Compulsive behavior. |
Internet—Social aspects. | Substance abuse.
Classification: LCC BF515 .L46 2020 (print) | LCC BF515 (ebook) |
DDC 152.4/2—dc23
LC record available at https://lccn.loc.gov/2020041077
LC ebook record available at https://lccn.loc.gov/2020041078

Dutton trade paperback ISBN: 9781524746742

Printed in the United States of America

8th Printing

BOOK DESIGN BY LORIE PAGNOZZI

For Mary, James, Elizabeth,
Peter, and little Lucas

CONTENTS

DOPAMINE NATION

The Problem

Feelin' good, feelin' good, all the money in
the world spent on feelin' good.

—J. B. LENOIR

This book is about pleasure. It's also about pain. Most important, it's about the relationship between pleasure and pain, and how understanding that relationship has become essential for a life well lived.

Why?

Because we've transformed the world from a place of scarcity to a place of overwhelming abundance: Drugs, food, news, gambling, shopping, gaming, texting, sexting, Facebooking, Instagramming, YouTubing, tweeting . . . the increased numbers, variety, and potency of highly rewarding stimuli today is staggering. The smartphone is the modern-day hypodermic needle, delivering digital dopamine 24/7 for a wired generation. If you haven't met your drug of choice yet, it's coming soon to a website near you.

Scientists rely on dopamine as a kind of universal currency for measuring the addictive potential of any experience. The more dopamine in the brain's reward pathway, the more addictive the experience.

In addition to the discovery of dopamine, one of the most remarkable neuroscientific findings in the past century is that the brain processes pleasure and pain in the same place. Further, pleasure and pain work like opposite sides of a balance.

We've all experienced that moment of craving a second piece of chocolate, or wanting a good book, movie, or video game to last forever. That moment of wanting is the brain's pleasure balance tipped to the side of pain.

This book aims to unpack the neuroscience of reward and, in so doing, enable us to find a better, healthier balance between pleasure and pain. But neuroscience is not enough. We also need the lived experience of human beings. Who better to teach us how to overcome compulsive overconsumption than those most vulnerable to it: people with addiction.

This book is based on true stories of my patients falling prey to addiction and finding their way out again. They've given me permission to tell their stories so that you might benefit from their wisdom, as I have. You may find some of these stories shocking, but to me they are just extreme versions of what we are all capable of. As philosopher and theologian Kent Dunnington wrote, "Persons with severe addictions are among those contemporary prophets that we ignore to our own demise, for they show us who we truly are."

Whether it's sugar or shopping, voyeuring or vaping, social media posts or *The Washington Post*, we all engage in behaviors we wish we didn't, or to an extent we regret. This book offers practical solutions for how to manage compulsive overconsumption in a world where consumption has become the all-encompassing motive of our lives.

In essence, the secret to finding balance is combining the science of desire with the wisdom of recovery.

PART I

The Pursuit of Pleasure

CHAPTER 1

Our Masturbation Machines

I went to greet Jacob in the waiting room. First impression? Kind. He was in his early sixties, middleweight, face soft but handsome . . . aging well enough. He wore the standard-issue Silicon Valley uniform: khakis and a casual button-down shirt. He looked unremarkable. Not like someone with secrets.

As Jacob followed me through the short maze of hallways, I could feel his anxiety like waves rolling off my back. I remembered when I used to get anxious walking patients back to my office. *Am I walking too fast? Am I swinging my hips? Does my ass look funny?*

It seems so long ago now. I admit I'm a battle-hardened version of my former self, more stoic, possibly more indifferent. *Was I a better doctor then, when I knew less and felt more?*

We arrived at my office and I shut the door behind him. Gently, I offered him one of two identical, equal-in-height, two-feet-apart, green-cushioned, therapy-sanctioned chairs. He sat. So did I. His eyes took in the room.

My office is ten by fourteen feet, with two windows, a desk with a computer, a sideboard covered with books, and a low table between the chairs. The desk, the sideboard, and the low table are all made of matching reddish-brown wood. The desk is a hand-me-down from my former department chair. It's cracked down the middle on the inside, where no one else can see it, an apt metaphor for the work I do.

On top of the desk are ten separate piles of paper, perfectly aligned, like an accordion. I am told this gives the appearance of organized efficiency.

The wall décor is a hodgepodge. The requisite diplomas, mostly unframed. Too lazy. A drawing of a cat I found in my neighbor's garbage, which I took for the frame but kept for the cat. A multicolored tapestry of children playing in and around pagodas, a relic from my time teaching English in China in my twenties. The tapestry has a coffee stain, but it's only visible if you know what you're looking for, like a Rorschach.

On display is an assortment of knickknacks, mostly gifts from patients and students. There are books, poems, essays, artwork, postcards, holiday cards, letters, cartoons.

One patient, a gifted artist and musician, gave me a photograph he had taken of the Golden Gate Bridge overlaid with his hand-drawn musical notes. He was no longer suicidal when he made it, yet it's a mournful image, all grays and blacks. Another patient, a beautiful young woman embarrassed by wrinkles that only she saw and no amount of Botox could erase, gave me a clay water pitcher big enough to serve ten.

To the left of my computer, I keep a small print of Albrecht Dürer's *Melencolia 1*. In the drawing, Melancholia personified as a woman sits stooped on a bench surrounded by the neglected tools of industry and time: a caliper, a scale, an hourglass, a hammer. Her starving dog, ribs protruding from his sunken frame, waits patiently and in vain for her to rouse herself.

To the right of my computer, a five-inch clay angel with wings wrought from wire stretches her arms skyward. The word *courage* is engraved at her feet. She's a gift from a colleague who was cleaning out her office. A leftover angel. I'll take it.

I'm grateful for this room of my own. Here, I am suspended out of time, existing in a world of secrets and dreams. But the space is also tinged with sadness and longing. When my patients leave my care, professional boundaries forbid that I contact them.

As real as our relationships are inside my office, they cannot exist outside this space. If I see my patients at the grocery store, I'm hesitant even to say hello lest I declare myself a human being with needs of my own. What, me eat?

Years ago, when I was in my psychiatry residency training, I saw my psychotherapy supervisor outside his office for the first time. He emerged from a shop wearing a trench coat and an Indiana Jones–style fedora. He looked like he'd just stepped off the cover of a J. Peterman catalogue. The experience was jarring.

I'd shared many intimate details of my life with him, and he had counseled me as he would a patient. I had not thought of him as a hat person. To me, it suggested a preoccupation with personal appearance that was at odds with the idealized version I had of him. But most of all, it made me aware of how disconcerting it might be for my own patients to see me outside my office.

I turned to Jacob and began. "What can I help you with?"

Other beginnings I've evolved over time include: "Tell me why you're here," "What brings you in today?" and even "Start at the beginning, wherever that is for you."

Jacob looked me over. "I am hoping," he said in a thick Eastern European accent, "you would be a man."

I knew then we would be talking about sex.

"Why?" I asked, feigning ignorance.

"Because it might be hard for you, a woman, to hear about my problems."

"I can assure you I've heard almost everything there is to hear."

"You see," he stumbled, looking shyly at me, "I have the sex addiction."

I nodded and settled into my chair. "Go on . . ."

Every patient is an unopened package, an unread novel, an unexplored land. A patient once described to me how rock climbing feels: When he's on the wall, nothing exists but infinite rock face juxtaposed against the finite decision of where next to put each finger and toe. Practicing psychotherapy is not unlike rock climbing. I immerse myself in story, the telling and retelling, and the rest falls away.

I've heard many variations on the tales of human suffering, but Jacob's story shocked me. What disturbed me most was what it implied about the world we live in now, the world we're leaving to our children.

Jacob started right in with a childhood memory. No pre-amble. Freud would have been proud.

"I masturbated first time when I was two or three years old," he said. The memory was vivid for him. I could see it on his face.

"I am on the moon," he continued, "but it is not really the moon. There is a person there like a god . . . and I have sexual experience which I don't recognize . . ."

I took *moon* to mean something like the abyss, nowhere and everywhere simultaneously. But what of God? Aren't we all yearning for something beyond ourselves?

As a young schoolboy, Jacob was a dreamer: buttons out of order, chalk on his hands and sleeves, the first to look out the window during lessons, and the last to leave the classroom for the day. He masturbated regularly by the time he was eight years old. Sometimes alone, sometimes with his best friend. They had not yet learned to be ashamed.

But after his First Communion, he was awakened to the idea of masturbation as a "mortal sin." From then on, he only masturbated alone, and he visited the Catholic priest of his family's local church every Friday to confess.

"I masturbate," he whispered through the latticed opening of the confessional.

"How many times?" asked the priest.

"Every day."

Pause. "Don't do it again."

Jacob stopped talking and looked at me. We shared a small smile of understanding. If such straightforward admonitions solved the problem, I would be out of a job.

Jacob the boy was determined to obey, to be "good," and so he clenched his fists and didn't touch himself there. But his resolve only ever lasted two or three days.

"That," he said, "was the beginning of my double life."

The term *double life* is as familiar to me as *ST segment elevation* is to the cardiologist, *stage IV* is to the oncologist, and *hemoglobin A1C* is to the endocrinologist. It refers to the addicted person's secret engagement with drugs, alcohol, or other compulsive behaviors, hidden from view, even in some cases from their own.

Throughout his teens, Jacob returned from school, went to the attic, and masturbated to a drawing of the Greek goddess Aphrodite he had copied from a textbook and hidden between the wooden floorboards. He would later look on this period of his life as a time of innocence.

At eighteen he moved to live with his older sister in the city to study physics and engineering at the university there. His sister was gone much of the day working, and for the first time in his life, he was alone for long stretches. He was lonely.

"So I decided to make a machine . . ."

"A machine?" I asked, sitting up a little straighter.

"A masturbation machine."

I hesitated. "I see. How did it work?"

"I connect a metal rod to a record player. The other end I connect to an open metal coil, which I wrap with a soft cloth." He drew a picture to show me.

"I put the cloth and the coil around my penis," he said, pronouncing *penis* as if it were two words: *pen* like the writing instrument, and *ness* like the Loch Ness Monster.

I had an urge to laugh but, after a moment's reflection, realized the urge was a cover for something else: I was afraid. Afraid that after inviting him to reveal himself to me, I wouldn't be able to help him.

"As the record player move round and round," he said, "the coil go up and down. I adjust the speed of the coil by adjusting the speed of the record player. I have three different speeds. In this way, I bring myself to the edge . . . many times, without going over. I also learn that smoking a cigarette at the same time brings me back from the edge, so I use this trick."

Through this method of microadjustments, Jacob was able to maintain a preorgasm state for hours. "This," he said, nodding, "very addictive."

Jacob masturbated for several hours a day using his machine. The pleasure for him was unrivaled. He swore he would stop. He hid the machine high up in a closet or dismantled it completely and threw away the parts. But a day or two later, he was pulling the parts down from the closet or out of the trash can, only to reassemble them and start again.

Perhaps you are repulsed by Jacob's masturbation machine, as I was when I first heard about it. Perhaps you regard it as a kind of extreme perversion that is beyond everyday experience, with little or no relevance to you and your life.

But if we do that, you and I, we miss an opportunity to appreciate something crucial about the way we live now: We are all, of a sort, engaged with our own masturbation machines.

Circa age forty, I developed an unhealthy attachment to romance novels. *Twilight*, a paranormal romance about teenage vampires, was my gateway drug. I was embarrassed enough to be reading it, much less admitting I was enthralled by it.

Twilight hit that sweet spot between love story, thriller, and fantasy, the perfect escape as I rounded the corner of my midlife bend. I was not alone. Millions of women my age were reading and fanning *Twilight*. There was nothing unusual per se about my getting caught up in a book. I've been a reader all my life. What was different was what happened next. Something I couldn't account for based on past proclivities or life circumstance.

When I finished *Twilight*, I ripped through every vampire romance I could get my hands on, and then moved on to werewolves, fairies, witches, necromancers, time travelers, soothsayers, mind readers, fire wielders, fortune-tellers, gem workers . . . you get the idea. At some point, tame love stories no longer satisfied, so I searched out increasingly graphic and erotic renditions of the classic boy-meets-girl fantasy.

I remember being shocked at how easy it was to find graphic sex scenes right there on the general fiction shelves at my neighborhood library. I worried that my kids had access to these books. The raciest thing at my local library growing up in the Midwest was *Are You There, God? It's Me, Margaret*.

Things escalated when, at the urging of my tech-savvier friend, I got a Kindle. No more waiting for books to be delivered from another library branch or hiding steamy book jackets behind medical journals, especially when my husband and kids were around. Now, with two swipes and a click, I had any book I wanted instantly, anywhere, anytime: on the train, on a plane, waiting to get my hair cut. I could just as easily pass off *Darkfever*, by Karen Marie Moning, as *Crime and Punishment* by Dostoyevsky.

In short, I became a chain reader of formulaic erotic genre novels. As soon as I finished one e-book, I moved on to the next: reading instead of socializing, reading instead of cooking, reading instead of sleeping, reading instead of paying attention to my husband and my kids. Once, I'm ashamed to admit, I brought my Kindle to work and read between patients.

I looked for ever-cheaper options all the way down to free. Amazon, like any good drug dealer, knows the value of a free sample. Once in a while I found a book of real quality that happened to be cheap; but most of the time, they were truly terrible, relying on worn-out plot devices and lifeless characters, chock-full of typos and grammatical errors. But I read them anyway because I was increasingly looking for a very specific type of experience. How I got there mattered less and less.

I wanted to indulge in that moment of mounting sexual tension that finally gets resolved when the hero and heroine hook up. I no longer cared about syntax, style, scene, or character. I just wanted my fix, and these books, written according to a formula, were designed to hook me.

Every chapter ended on a note of suspense, and the chapters themselves built toward the climax. I started rushing through the first part of the book until I got to the climax and didn't bother to read the rest after it was done. I am now sadly in possession of the knowledge that if you open any romance novel to approximately three-quarters of the way through, you can get right to the point.

About a year into my new obsession with romance, I found myself up at 2:00 a.m. on a weeknight reading *Fifty Shades of Grey*. I rationalized it was a modern-day telling of *Pride and Prejudice*—right up until I got to the page on "butt plugs" and had a flash of insight that reading about sadomasochistic sex toys in the wee hours of the morning was not how I wanted to be spending my time.

Addiction broadly defined is the continued and compulsive consumption of a substance or behavior (gambling, gaming, sex) despite its harm to self and/or others.

What happened to me is trivial compared to the lives of those with overpowering addiction, but it speaks to the growing problem of compulsive overconsumption that we all face today, even when our lives are good. I have a kind and loving husband, great kids, meaningful work, freedom, autonomy, and relative wealth—no trauma, social dislocation, poverty, unemployment, or other risk factors for addiction. Yet I was compulsively retreating further and further into a fantasy world.

The Dark Side of Capitalism

At age twenty-three, Jacob met and married his wife. They moved together into the three-room apartment she shared with her parents, and he left his machine behind—forever, he hoped. He and his wife registered to get an apartment of their own but were told the wait would be twenty-five years. This was typical in the 1980s in the Eastern European country where they lived.

Instead of consigning themselves to decades of living with her parents, they decided to earn extra money on the side to buy their own place sooner. They started a computer business importing machines from Taiwan, joining the growing underground economy.

Their business prospered, and they soon became rich by local standards. They acquired a house and plot of land. They had two children, a son and a daughter.

Their upward trajectory seemed assured when Jacob was offered a job working as a scientist in Germany. They jumped at the chance to move west, further his career, and provide their children with all the opportunities that Western Europe could offer. The move offered opportunities all right, not all of them good.

"Once we move to Germany, I discover pornography, porn-kinos, live shows. This town I live in is known for this, and I cannot resist. But I manage. I manage for ten years. I am working as a scientist, working hard, but in 1995, everything change."

"What changed?" I asked, already guessing the answer.

"The Internet. I am forty-two years old, and doing okay, but with the Internet, my life start to fall apart. Once in 1999, I am in same hotel room I stay in maybe fifty times before. I have big conference, big talk the next day. But I stay up all night watching porn instead of preparing my talk. I show up at the conference with no sleep and no talk. I give a speech, very bad. I almost lose my job." He looked down and shook his head, remembering.

"After that I start a new ritual," he said. "Every time I go into hotel room, I place sticky notes all around—on the bathroom mirror, the TV, the remote control—saying, 'Don't do it.' I don't even last one day."

I was struck by how much hotel rooms are like latter-day Skinner boxes: a bed, a TV, and a minibar. Nothing to do but press the lever for drug.

He looked down again and the silence stretched. I gave him time.

"That was when I first think about ending my life. I think the world will not miss me, and maybe better without me. I walk to the balcony and look down. Four stories . . . that would be enough."

———

One of the biggest risk factors for getting addicted to any drug is easy access to that drug. When it's easier to get a drug, we're more likely to try it. In trying it, we're more likely to get addicted to it.

The current US opioid epidemic is a tragic and compelling example of this fact. The quadrupling of opioid prescribing (OxyContin, Vicodin, Duragesic fentanyl) in the United States between 1999 and 2012, combined with widespread distribution of those opioids to every corner of America, led to rising rates of opioid addiction and related deaths.

A task force appointed by the Association of Schools and Programs of Public Health (ASPPH) issued a report on November 1, 2019, concluding, "The tremendous expansion of the supply of powerful (high-potency as well as long-acting) prescription opioids led to scaled increases in prescription opioid dependence, and the transition of many to illicit opioids, including fentanyl and its analogs, which have subsequently driven exponential increases in overdose." The report also stated that opioid use disorder "is caused by repeated exposure to opioids."

Likewise, decreasing the supply of addictive substances decreases exposure and risk of addiction and related harms. A natural experiment in the last century to test and prove this hypothesis was Prohibition, a nationwide constitutional ban on the production, importation, transportation, and sale of alcoholic beverages in the United States from 1920 to 1933.

Prohibition led to a sharp decrease in the number of Americans consuming and becoming addicted to alcohol. Rates of public drunkenness and alcohol-related liver disease decreased by half during this period in the absence of new remedies to treat addiction.

There were unintended consequences, of course, such as the creation of a large black market run by criminal gangs. But the positive impact of Prohibition on alcohol consumption and related morbidity is widely underrecognized.

The reduced drinking effects of Prohibition persisted through the 1950s. Over the subsequent thirty years, as alcohol became more available again, consumption steadily increased.

In the 1990s, the percentage of Americans who drank alcohol increased by almost 50 percent, while high-risk drinking increased by 15 percent. Between 2002 and 2013, diagnosable alcohol addiction rose by 50 percent in older adults (over age sixty-five) and 84 percent in women, two demographic groups who had previously been relatively immune to this problem.

To be sure, increased access is not the only risk for addiction. The risk increases if we have a biological parent or grandparent with addiction, even when we're raised outside the addicted home. Mental illness is a risk factor, although the relationship between the two is unclear: Does the mental illness lead to drug use, does drug use cause or unmask mental illness, or is it somewhere in between?

Trauma, social upheaval, and poverty contribute to addiction risk, as drugs become a means of coping and lead to epigenetic changes—heritable changes to the strands of DNA outside of inherited base pairs—affecting gene expression in both an individual and their offspring.

These risk factors notwithstanding, increased access to addictive substances may be the most important risk factor facing modern people. Supply has created demand as we all fall prey to the vortex of compulsive overuse.

Our dopamine economy, or what historian David Courtwright has called "limbic capitalism," is driving this change, aided by transformational technology that has increased not just access but also drug numbers, variety, and potency.

The cigarette-rolling machine invented in 1880, for example, made it possible to go from four cigarettes rolled per minute to a staggering 20,000. Today, 6.5 trillion cigarettes are sold annually around the world, translating to roughly 18 billion cigarettes consumed per day, responsible for an estimated 6 million deaths worldwide.

In 1805, the German Friedrich Sertürner, while working as a pharmacist's apprentice, discovered the painkiller morphine—an opioid alkaloid ten times more potent than its precursor opium. In 1853, the Scottish physician Alexander Wood invented the hypodermic syringe. Both of these inventions contributed to hundreds of reports in late-nineteenth-century medical journals of iatrogenic (physician-initiated) cases of morphine addiction.

In an attempt to find a less addictive opioid painkiller to replace morphine, chemists came up with a brand-new compound, which they named "heroin" for *heroisch*, the German word for "courageous." Heroin turned out to be two to five times more potent than morphine and gave way to the narcomania of the early 1900s.

Today, potent pharmaceutical-grade opioids such as oxycodone, hydrocodone, and hydromorphone are available in every imaginable form: pills, injection, patch, nasal spray. In 2014, a middle-aged patient walked into my office sucking on a bright red fentanyl lollipop. Fentanyl, a synthetic opioid, is fifty to one hundred times more potent than morphine.

Beyond opioids, many other drugs are also more potent today than in yesteryear. Electronic cigarettes—chic, discreet, odorless, rechargeable nicotine delivery systems—lead to higher levels of blood nicotine over shorter periods of consumption than traditional cigarettes. They also come in a multitude of flavors designed to appeal to teenagers.

Today's cannabis is five to ten times more potent than the cannabis of the 1960s and is available in cookies, cakes, brownies, gummy bears, blueberries, "pot tarts," lozenges, oils, aromatics, tinctures, teas . . . the list is endless.

Food is manipulated by technicians around the world. Following World War I, the automation of chip and fry production lines led to the creation of the bagged potato chip. In 2014, Americans consumed 112.1 pounds of potatoes per person, of which 33.5 pounds were fresh potatoes and the remaining 78.5 pounds were processed. Copious amounts of sugar, salt, and fat are added to much of the food we eat, as well as thousands of artificial flavors to satisfy our modern appetite for things like French toast ice cream and Thai tomato coconut bisque.

With increasing access and potency, polypharmacy—that is, using multiple drugs simultaneously or in close proximity—has become the norm. My patient Max found it easier to draw out a timeline of his drug use than to explain it to me.

As you can see in his illustration, he started at age seventeen with alcohol, cigarettes, and cannabis ("Mary Jane"). By age eighteen, he was snorting cocaine. At age nineteen, he switched to OxyContin and Xanax. Through his twenties, he used Percocet, fentanyl, ketamine, LSD, PCP, DXM, and MXE, eventually landing on Opana, a pharmaceutical-grade opioid that led him to heroin, where he stayed until he came to see me at age thirty. In total, he went through fourteen different drugs in a little over a decade.

DRUG USE TIMELINE

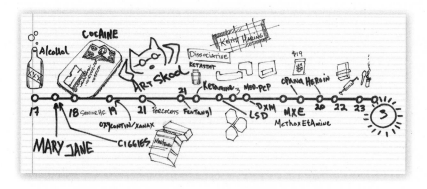

The world now offers a full complement of digital drugs that didn't exist before, or if they did exist, they now exist on digital platforms that have exponentially increased their potency and availability. These include online pornography, gambling, and video games, to name a few.

Furthermore, the technology itself is addictive, with its flashing lights, musical fanfare, bottomless bowls, and the promise, with ongoing engagement, of ever-greater rewards.

My own progression from a relatively tame vampire romance novel to what amounts to socially sanctioned pornography for women can be traced to the advent of the electronic reader.

The act of consumption itself has become a drug. My patient Chi, a Vietnamese immigrant, got hooked on the cycle of searching for and buying products online. The high for him began with deciding what to buy, continued through anticipating delivery, and culminated in the moment he opened the package.

Unfortunately, the high didn't last much beyond the time it took him to rip off the Amazon tape and see what was inside. He had rooms full of cheap consumer goods and was tens of thousands of dollars in debt. Even then, he couldn't stop. To keep the cycle going, he resorted to ordering ever-cheaper goods—key chains, mugs, plastic sunglasses—and returning them immediately upon arrival.

The Internet and Social Contagion

Jacob decided not to end his life that day in the hotel. The very next week, his wife was diagnosed with brain cancer. They returned to their home country and he spent the next three years taking care of her until she died.

In 2001, at age forty-nine, he reconnected with and married his high school sweetheart.

"I tell her before we marry about my problem. But maybe I minimize when I tell her."

Jacob and his new wife bought a home together in Seattle. Jacob commuted to a job as a scientist in Silicon Valley. The more time he spent in Silicon Valley and away from his wife, the more he returned to old patterns of pornography and compulsive masturbation.

"I never do pornography when we are together. But when I am here in Silicon Valley or traveling, and she is not with me, then I do."

Jacob paused. What came next was clearly difficult for him to talk about.

"Sometimes when I play with electricity, in my job, I can

feel something in my hands. I am curious. I begin to wonder what it would feel like to touch my penis with a current. So I start to research online, and I discover a whole community of people using electrical stimulation.

"I attach electrodes and wires to my stereo system. I try an alternating current using the voltage from the stereo system. Then instead of simple wire, I attach electrodes made of cotton in salty water. The higher the volume on the stereo, the higher the current. At low volume, I feel nothing. At higher volume, it is painful. In between, I can orgasm from the sensation."

My eyes got wide. I couldn't help it.

"But this very dangerous," he continued. "I realize if a power outage, this could lead to power surge, and then I could get hurt. People have died doing this. Online I learn I can buy a medical kit, like a . . . what do you call them, those machines to treat pain . . ."

"A TENS unit?"

"Yes, a TENS unit, for six hundred dollars, or I can make my own for twenty dollars. I decide to make my own. I buy the material. I make the machine. It works. It works well." He paused. "But then the real discovery. I can program it. I can create custom routines and synchronize the music with the feeling."

"What kinds of routines?"

"Hand job, blow job. You name it. And then I discover not just my routines. I go online and download other people's routines, and share mine. Some people write programs to sync up with porn videos, so you feel what you're watching . . . just like virtual reality. The pleasure, it comes from the sensation of course, but also from building the machine, and anticipating what it will do, and experimenting with ways to improve it, and sharing with others."

He smiled, remembering, just before his face fell, anticipating what came next. Scrutinizing me, I could tell he was gauging whether I could take it. I braced myself and nodded for him to go on.

"It gets worse. There are chat rooms where you can watch people pleasure themselves, live. It's free to watch, but option to buy tokens. I give tokens for good performance. I film myself and put online. Just my private parts. No other part of me. It is exhilarating at first, having strangers watch me. But I feel guilty too, that watching would give others the idea, and they might get addicted."

In 2018, I served as a medical expert witness in the case of a man who plowed his truck into two teenagers, killing both. He was driving under the influence of drugs. As part of that litigation, I spent time talking to Detective Vince Dutto, a lead crime investigator in Placer County, California, where the trial took place.

Curious about his work, I asked him about any changes in patterns he'd seen over the last twenty years. He told me about the tragic case of a six-year-old boy who sodomized his younger, four-year-old brother.

"Normally, when we get these calls," he said, "it's because some adult the child has contact with is sexually abusing him, and then the kid reenacts it on another kid, like his little brother. But we did a thorough investigation and there was no evidence the older brother was being abused. His parents were divorced and worked a lot, so the kids were kind of raising themselves, but there was no active sexual abuse going on.

"What eventually came out in this case was the older brother had been watching cartoons on the Internet and stumbled across some Japanese anime cartoon showing all kinds of sex acts. The kid had his own iPad, and no one was policing what he was doing, and after watching a bunch of these cartoons, he decided to try it out on his little brother. Now, that kind of thing, in more than twenty years of police work, I've never seen before."

The Internet promotes compulsive overconsumption not merely by providing increased access to drugs old and new, but also by suggesting behaviors that otherwise may never have occurred to us. Videos don't just "go viral." They're literally contagious, hence the advent of the meme.

Human beings are social animals. When we see others behaving in a certain way online, those behaviors seem "normal" because other people are doing them. "Twitter" is an apt name for the social media messaging platform favored by pundits and presidents alike. We are like flocks of birds. No sooner has one of us raised a wing in flight than the entire flock of us is rising into the air.

***

Jacob looked down at his hands. He couldn't meet my eyes.

"Then I meet a lady in this chat room. She like to dominate

men. I introduce her into the electrical stuff, and then I give her the ability to control the electricity remotely: frequency, volume, structure of the pulses. She likes to bring me to the edge, and then let me not go over. She does this ten times, and other people watch, and make comments. We develop the friendship, this lady and I. She never wants to show her face. But I saw her once, by accident, when her camera fell for a moment."

"How old was she?" I asked.

"In her forties, I guess . . ."

I wanted to ask what she looked like but sensed my own prurient curiosity at play here, rather than his therapeutic needs, so I refrained.

Jacob said, "My wife discover all this, and she say she will leave me. I promise to stop. I tell my lady friend online I am quitting. My lady friend very angry. My wife very angry. I hate myself then. I stop for a while. Maybe a month. But then I start up again. Just me and my machine, not the chat rooms. I lie to my wife, but eventually she discover. Her therapist tell her to leave me. So my wife, she leave me. She move to our house in Seattle, and now I am alone."

Shaking his head, he said, "It never as good as I imagine. The reality always less. I tell myself never again, and I destroy the machine and throw it away. But at four a.m. the next morning, I am getting it from the trash and building it again."

Jacob looked at me with pleading eyes. "I want to stop. I want to. I don't want to die an addict."

I'm not sure what to say. I imagine him attached by his genitals through the Internet to a room full of strangers. I feel horror, compassion, and a vague and disquieting sense that it could have been me.

———

Not unlike Jacob, we are all at risk of titillating ourselves to death.

Seventy percent of world global deaths are attributable to modifiable behavioral risk factors like smoking, physical inactivity, and diet. The leading global risks for mortality are high blood pressure (13 percent), tobacco use (9 percent), high blood sugar (6 percent), physical inactivity (6 percent), and obesity (5 percent). In 2013, an estimated 2.1 billion adults were overweight, compared with 857 million in 1980. There are now more people worldwide, except in parts of sub-Saharan Africa and Asia, who are obese than who are underweight.

Rates of addiction are rising the world over. The disease burden attributed to alcohol and illicit drug addiction is 1.5 percent globally, and more than 5 percent in the United States. These data exclude tobacco consumption. Drug of choice varies by country. The US is dominated by illicit drugs, Russia and Eastern Europe by alcohol addiction.

Global deaths from addiction have risen in all age groups between 1990 and 2017, with more than half the deaths occurring in people younger than fifty years of age.

The poor and undereducated, especially those living in rich nations, are most susceptible to the problem of compulsive overconsumption. They have easy access to high-reward, high-potency, high-novelty drugs at the same time that they lack access to meaningful work, safe housing, quality education, affordable health care, and race and class equality before the law. This creates a dangerous nexus of addiction risk.

Princeton economists Anne Case and Angus Deaton have shown that middle-aged white Americans without a college degree are dying younger than their parents, grandparents, and great-grandparents. The top three leading causes of death in this group are drug overdoses, alcohol-related liver disease, and suicides. Case and Deaton have aptly called this phenomenon "deaths of despair."

Our compulsive overconsumption risks not just our demise but also that of our planet. The world's natural resources are rapidly diminishing. Economists estimate that in 2040 the world's natural capital (land, forests, fisheries, fuels) will be 21 percent less in high-income countries and 17 percent less in poorer countries than today. Meanwhile, carbon emissions will grow by 7 percent in high-income countries and 44 percent in the rest of the world.

We are devouring ourselves.

CHAPTER 2

Running from Pain

I met David in 2018. He was physically unremarkable: white, medium build, brown hair. He had an uncertainty about him that made him seem younger than the thirty-five years documented in the medical record. I found myself thinking, *He won't last. He'll come back to clinic once or twice and I'll never see him again.*

But I've learned my powers of prognostication are unreliable. I've had patients I was convinced I could help who proved to be intractable, and others I deemed hopeless who were surprisingly resilient. Hence, when seeing new patients now, I try to quiet that doubting voice and remember that everyone's got a shot at recovery.

"Tell me what brings you in," I said.

David's problems began in college, but more precisely the day he walked into student mental health services. He was a twenty-year-old sophomore undergraduate in upstate New York looking for help with anxiety and poor school performance.

His anxiety was triggered by interacting with strangers, or anyone he didn't know well. His face would flush, his chest and back would get damp, and his thoughts would get jumbled. He avoided classes where he had to speak in front of others. He dropped out of a required speech and communications seminar twice, eventually fulfilling the requirement by taking an equivalent class at community college.

"What were you afraid of?" I asked.

"I was afraid to fail. I was afraid to be exposed as not knowing. I was afraid to ask for help."

After a forty-five-minute appointment and a pencil-and-paper test that took less than five minutes to complete, he was diagnosed with attention deficit disorder (ADD) and generalized anxiety disorder (GAD). The psychologist who administered the test recommended he follow up with a psychiatrist to prescribe an antianxiety medication and, David said, a "stimulant for my ADD." He was not offered psychotherapy or other nonmedication behavioral modification.

David went to see a psychiatrist, who prescribed Paxil, a selective serotonin reuptake inhibitor to treat depression and anxiety, and Adderall, a stimulant to treat ADD.

"So how did it go for you—the meds, I mean?"

"The Paxil helped with the anxiety a little at first. It dampened down some of the worst sweating, but it wasn't a cure. I ended up changing my major from computer engineering to computer science, thinking that would help. It required less interaction.

"But because I wasn't able to speak up and say I didn't know, I failed an exam. Then I failed the next one. Then I dropped out for a semester not to take a hit on my grade point average. Eventually, I switched out of the school of engineering altogether, which was really sad because it was what I loved and really wanted to do. I became a history major: The classes were smaller, only twenty people, and I could get away with being less interactive. I could take the blue book home and work by myself."

"What about the Adderall?" I asked.

"I'd take ten milligrams first thing every morning before class. It helped me get that deep focus. But looking back, I think I just had bad study habits. Adderall helped me make up for that, but it also helped me procrastinate. If there was a test and I hadn't studied, I'd take Adderall around the clock, all through the day and night, to cram for the exam. Then it got to where I couldn't study without it. Then I started needing more."

"Was it hard to get more?"

"Not really," he said. "I always knew when a refill was due. I'd call the psychiatrist a few days before. Not a lot of days before, just one or two, so they wouldn't get suspicious. Actually, I'd run out like . . . ten days before, but if I called a few days before, they'd refill it right then. I also learned it was better to talk to the P.A., the physician's assistant. They'd be more likely to refill without asking too many questions. Sometimes I'd make up excuses, like say there was a problem with the mail-order pharmacy. But most of the time I didn't have to."

"It sounds like the pills weren't really helping."

David paused. "In the end, it came down to comfort. It was easier to take a pill than feel the pain."

In 2016, I gave a presentation on drug and alcohol problems to faculty and staff at the Stanford student mental health clinic. It had been some months since I'd been to that part of campus. I arrived early and, while I waited in the front lobby to meet my contact, my attention was drawn to a wall of brochures for the taking.

There were four brochures in all, each with some variation of the word *happiness* in the title: *The Habit of Happiness*, *Sleep Your Way to Happiness*, *Happiness Within Reach*, and *7 Days to a Happier You*. Inside each brochure were prescriptions for achieving happiness: "List 50 things that make you happy," "Look at yourself in the mirror [and] list things you love about yourself in your journal," and "Produce a stream of positive emotions."

Perhaps most telling of all: "Optimize timing and variety of happiness strategies. Be intentional about when and how often. For acts of kindness: Self-experiment to determine whether performing many good deeds in one day or one act each day is most effective for you."

These brochures illustrate how the pursuit of personal happiness has become a modern maxim, crowding out other definitions of the "good life." Even acts of kindness toward others are framed as a strategy for personal happiness. Altruism, no longer merely a good in itself, has become a vehicle for our own "well-being."

Philip Rieff, a mid-twentieth-century psychologist and philosopher, foresaw this trend in *The Triumph of the Therapeutic*:

Uses of Faith After Freud: "Religious man was born to be saved; psychological man is born to be pleased."

Messages exhorting us to pursue happiness are not limited to the realm of psychology. Modern religion too promotes a theology of self-awareness, self-expression, and self-realization as the highest good.

In his book *Bad Religion*, writer and religious scholar Ross Douthat describes our New Age "God Within" theology as "a faith that's at once cosmopolitan and comforting, promising all the pleasures of exoticism . . . without any of the pain . . . a mystical pantheism, in which God is an experience rather than a person. . . . It's startling how little moral exhortation there is in the pages of the God Within literature. There are frequent calls to 'compassion' and 'kindness,' but little guidance for people facing actual dilemmas. And what guidance there is often amounts to 'if it feels good, do it.'"

My patient Kevin, nineteen years old, was brought to see me by his parents in 2018. Their concerns were the following: He wouldn't go to school, couldn't keep a job, and wouldn't follow any of the household rules.

His parents were as imperfect as the rest of us, but they were trying hard to help him. There was no evidence of abuse or neglect. The problem was they seemed unable to put any constraints on him. They worried that by making demands, they would "stress him out" or "traumatize him."

Perceiving children as psychologically fragile is a quintessentially modern concept. In ancient times, children were considered miniature adults, fully formed from birth. For most of Western civilization, children were regarded as innately evil.

The job of parents and caregivers was to enforce extreme discipline in order to socialize them to live in the world. It was entirely acceptable to use corporal punishment and fear tactics to get a child to behave. No longer.

Today, many parents I see are terrified of doing or saying something that will leave their child with an emotional scar, thereby setting them up, so the thinking goes, for emotional suffering and even mental illness in later life.

This notion can be traced to Freud, whose groundbreaking psychoanalytic contribution was that early childhood experiences, even those long forgotten or outside of conscious awareness, can cause lasting psychological damage. Unfortunately, Freud's insight that early childhood trauma can influence adult psychopathology has morphed into the conviction that any and every challenging experience primes us for the psychotherapy couch.

Our efforts to insulate our children from adverse psychological experiences play out not just in the home but also in school. At the primary school level, every child receives some equivalent of the "Star of the Week" award—not for any particular accomplishment but in alphabetical order. Every child is taught to be on the lookout for bullies lest they become bystanders instead of upstanders. At the university level, faculty and students talk about triggers and safe spaces.

That parenting and education are informed by developmental psychology and empathy is a positive evolution. We should acknowledge every person's worth independent of achievement, stop physical and emotional brutality on the schoolyard and everywhere else, and create safe spaces to think, learn, and discuss.

But I worry that we have both oversanitized and over-pathologized childhood, raising our children in the equivalent of a padded cell, with no way to injure themselves but also no means to ready themselves for the world.

By protecting our children from adversity, have we made them deathly afraid of it? By bolstering their self-esteem with false praise and a lack of real-world consequences, have we made them less tolerant, more entitled, and ignorant of their own character defects? By giving in to their every desire, have we encouraged a new age of hedonism?

Kevin shared his life philosophy with me in one of our sessions. I must admit I was horrified.

"I do whatever I want, whenever I want. If I want to stay in my bed, I stay in my bed. If I want to play video games, I play video games. If I want to snort a line of coke, I text my dealer, he drops it off, and I snort a line of coke. If I want to have sex, I go online and find someone and meet them and have sex."

"How's that working out for you, Kevin?" I asked.

"Not very well." For a single instant he looked ashamed.

Over the past three decades, I have seen growing numbers of patients like David and Kevin who appear to have every advantage in life—supportive families, quality education, financial stability, good health—yet develop debilitating anxiety, depression, and physical pain. Not only are they not functioning to their potential; they're barely able to get out of bed in the morning.

The practice of medicine has likewise been transformed by our striving for a pain-free world.

Prior to the 1900s, doctors believed some degree of pain was healthy. Leading surgeons of the 1800s were reluctant to adopt general anesthesia during surgery because they believed that pain boosted the immune and cardiovascular response and expedited healing. Although there's no evidence I know of showing that pain in fact speeds up tissue repair, there is emerging evidence that taking opioids during surgery slows it down.

The famous seventeenth-century physician Thomas Sydenham said this about pain: "I look upon every . . . effort calculated totally to subdue that pain and inflammation dangerous in the extreme. . . . For certainty a moderate degree of pain and inflammation in the extremities are the instruments which nature makes use of for the wisest purposes."

By contrast, doctors today are expected to eliminate all pain lest they fail in their role as compassionate healers. Pain in any form is considered dangerous, not just because it hurts but also because it's thought to kindle the brain for future pain by leaving a neurological wound that never heals.

The paradigm shift around pain has translated into massive prescribing of feel-good pills. Today, more than one in four American adults—and more than one in twenty American children—takes a psychiatric drug on a daily basis.

The use of antidepressants like Paxil, Prozac, and Celexa is rising in countries all over the world, with the United States topping the list. Greater than one in ten Americans (110 people per 1,000) takes an antidepressant, followed by Iceland (106/1,000), Australia (89/1,000), Canada (86/1,000), Denmark (85/1,000), Sweden (79/1,000), and Portugal (78/1,000). Among twenty-five countries, South Korea was last (13/1,000).

Antidepressant use rose 46 percent in Germany in just four years, and 20 percent in Spain and Portugal during the same period. Although data for other Asian countries, including China, are not available, we can infer growing use of antidepressants by looking at sales trends. In China, sales of antidepressants reached $2.61 billion in 2011, up 19.5 percent from the previous year.

Prescriptions of stimulants (Adderall, Ritalin) in the United States doubled between 2006 and 2016, including in children younger than five years old. In 2011, two-thirds of American children diagnosed with ADD were prescribed a stimulant.

Prescriptions for sedative medications like benzodiazepines (Xanax, Klonopin, Valium), also addictive, are on the rise, perhaps to compensate for all those stimulants we're taking. Between 1996 and 2013 in the United States, the number of adults who filled a benzodiazepine prescription increased by 67 percent, from 8.1 million to 13.5 million people.

In 2012, enough opioids were prescribed for every American to have a bottle of pills, and opioid overdoses killed more Americans than guns or car accidents.

Is it any wonder, then, that David assumed he should numb himself with pills?

Beyond extreme examples of running from pain, we've lost the ability to tolerate even minor forms of discomfort. We're constantly seeking to distract ourselves from the present moment, to be entertained.

As Aldous Huxley said in *Brave New World Revisited*, "the development of a vast mass communications industry, concerned in the main neither with the true nor the false, but with the unreal, the more or less totally irrelevant . . . failed to take into account man's almost infinite appetite for distractions."

Along similar lines, Neil Postman, the author of the 1980s classic *Amusing Ourselves to Death*, wrote, "Americans no longer talk to each other, they entertain each other. They do not exchange ideas, they exchange images. They do not argue with propositions; they argue with good looks, celebrities, and commercials."

My patient Sophie, a Stanford undergraduate from South Korea, came in seeking help for depression and anxiety. Among the many things we talked about, she told me she spends most of her waking hours plugged into some kind of device: Instagramming, YouTubing, listening to podcasts and playlists.

In session with her I suggested she try walking to class without listening to anything and just letting her own thoughts bubble to the surface.

She looked at me both incredulous and afraid.

"Why would I do that?" she asked, openmouthed.

"Well," I ventured, "it's a way of becoming familiar with yourself. Of letting your experience unfold without trying to control it or run away from it. All that distracting yourself with devices may be contributing to your depression and anxiety. It's pretty exhausting avoiding yourself all the time. I wonder if experiencing yourself in a different way might give you access to new thoughts and feelings, and help you feel more connected to yourself, to others, and to the world."

She thought about that for a moment. "But it's so *boring*," she said.

"Yes, that's true," I said. "Boredom is not just boring. It can also be terrifying. It forces us to come face-to-face with bigger questions of meaning and purpose. But boredom is also an opportunity for discovery and invention. It creates the space necessary for a new thought to form, without which we're endlessly reacting to stimuli around us, rather than allowing ourselves to be within our lived experience."

The next week, Sophie experimented with walking to class without being plugged in.

"It was hard at first," she said. "But then I got used to it and even kind of liked it. I started noticing the trees."

Lack of Self-Care or Mental Illness?

Back to David, who was, in his own words, taking "Adderall around the clock." After he graduated from college in 2005, he moved back in with his parents. He thought about going to law school, took the LSATs, and even did okay, but when it came down to applying, he didn't feel like it.

"I mostly sat on the couch and built up a lot of anger and resentment: at myself, at the world."

"What were you angry about?"

"I felt like I'd wasted my undergraduate education. I hadn't studied what I really wanted to study. My girlfriend was still back at school . . . doing great, getting a master's. I was wallowing at home doing nothing."

After David's girlfriend graduated, she landed a job in Palo Alto. He followed her there, and in 2008 they were married. David got a job at a technology start-up, where he interacted with young, smart engineers who were generous with their time.

He got back into coding and learned all the stuff he had meant to study in college but was too afraid to pursue in a room full of students. He got promoted to software developer, was working fifteen-hour days, and ran thirty miles a week in his spare time.

"But to make all that happen," he said, "I was taking more Adderall, not just in the morning, but all through the day. I'd wake up in the morning, take Adderall. Get home, eat dinner, take more Adderall. Pills became my new normal. I was also drinking huge amounts of caffeine. Then I'd hit the end of the night, and I needed to go to sleep, and I'm like, *Okay, what do I do now?* So I went back to the psychiatrist and talked her into giving me Ambien. I pretended like I didn't know what Ambien was, but my mom had taken Ambien for a long time, and a couple uncles too. I also talked her into a limited prescription of Ativan for anxiety before presentations. From

2008 to 2018, I was taking up to thirty milligrams of Adderall a day, fifty milligrams of Ambien a day, and three to six milligrams of Ativan a day. I thought, *I have anxiety and ADHD and I need this to function.*"

David attributed fatigue and inattentiveness to a mental illness rather than to sleep deprivation and overstimulation, a logic he used to justify continued use of pills. I've seen a similar paradox in many of my patients over the years: They use drugs, prescribed or otherwise, to compensate for a basic lack of self-care, then attribute the costs to a mental illness, thus necessitating the need for more drugs. Hence poisons become vitamins.

"You were getting your A vitamins: Adderall, Ambien, and Ativan," I joked.

He smiled. "I guess you could say that."

"Did your wife or anybody else know what was going on with you?"

"No. Nobody did. My wife had no idea. Sometimes I would drink alcohol when I ran out of Ambien, or get angry and yell at her when I took too much Adderall. But other than that, I hid it pretty well."

"So then what happened?"

"I got tired of it. Tired of taking uppers and downers day and night. I started thinking about ending my life. I thought I'd be better off, and other people would be better off. But my wife was pregnant, so I knew I needed to make a change. I told her I needed help. I asked her to take me to the hospital."

"How did she react?"

"She took me to the emergency room, and when it all came out, she was shocked."

"What shocked her?"

"The pills. All the pills I was taking. My huge stash. And how much I had been hiding."

David was admitted to the inpatient psychiatric ward and diagnosed with stimulant and sedative addiction. He stayed in the hospital until he finished withdrawing from Adderall, Ambien, and Ativan, and until he was no longer suicidal. It took two weeks. He was discharged home to his pregnant wife.

———

We're all running from pain. Some of us take pills. Some of us couch surf while binge-watching Netflix. Some of us read romance novels. We'll do almost anything to distract ourselves from ourselves. Yet all this trying to insulate ourselves from pain seems only to have made our pain worse.

According to the World Happiness Report, which ranks 156 countries by how happy their citizens perceive themselves to be, people living in the United States reported being less happy in 2018 than they were in 2008. Other countries with similar measures of wealth, social support, and life expectancy saw similar decreases in self-reported happiness scores, including Belgium, Canada, Denmark, France, Japan, New Zealand, and Italy.

Researchers interviewed nearly 150,000 people in twenty-six countries to determine the prevalence of generalized anxiety disorder, defined as excessive and uncontrollable worry that adversely affected their life. They found that richer countries had higher rates of anxiety than poor ones. The authors wrote, "The disorder is significantly more prevalent and impairing in high-income countries than in low- or middle-income countries."

The number of new cases of depression worldwide increased 50 percent between 1990 and 2017. The highest increases in new cases were seen in regions with the highest sociodemographic index (income), especially North America.

Physical pain too is increasing. Over the course of my career, I have seen more patients, including otherwise healthy young people, presenting with full body pain despite the absence of any identifiable disease or tissue injury. The numbers and types of unexplained physical pain syndromes have grown: complex regional pain syndrome, fibromyalgia, interstitial cystitis, myofascial pain syndrome, pelvic pain syndrome, and so on.

When researchers asked the following question to people in thirty countries around the world—"During the past four weeks, how often have you had bodily aches or pains? Never; seldom; sometimes; often; or very often?"—they found that Americans reported more pain than any other country.

Thirty-four percent of Americans said they felt pain "often" or "very often," compared to 19 percent of people living in China, 18 percent of people living in Japan, 13 percent of people living in Switzerland, and 11 percent of people living in South Africa.

The question is: Why, in a time of unprecedented wealth, freedom, technological progress, and medical advancement, do we appear to be unhappier and in more pain than ever?

The reason we're all so miserable may be because we're working so hard to avoid being miserable.

CHAPTER 3

The Pleasure-Pain Balance

Neuroscientific advances in the last fifty to one hundred years, including advances in biochemistry, new imaging techniques, and the emergence of computational biology, shed light on fundamental reward processes. By better understanding the mechanisms that govern pain and pleasure, we can gain new insight into why and how too much pleasure leads to pain.

Dopamine

The main functional cells of the brain are called neurons. They communicate with each other at synapses via electrical signals and neurotransmitters.

Neurotransmitters are like baseballs. The pitcher is the presynaptic neuron. The catcher is the postsynaptic neuron. The space between pitcher and catcher is the synaptic cleft. Just as the ball is thrown between pitcher and catcher, neurotransmitters bridge the distance between neurons: chemical messengers regulating electrical signals in the brain.

There are many important neurotransmitters, but let's focus on dopamine.

NEUROTRANSMITTER

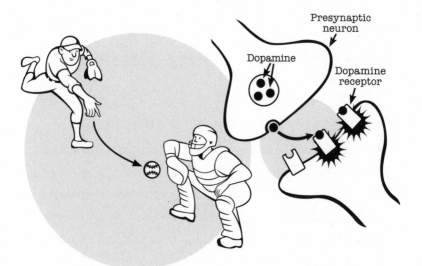

Dopamine was first identified as a neurotransmitter in the human brain in 1957 by two scientists working independently: Arvid Carlsson and his team in Lund, Sweden, and Kathleen Montagu, based outside of London. Carlsson went on to win the Nobel Prize in Physiology or Medicine.

Dopamine is not the only neurotransmitter involved in reward processing, but most neuroscientists agree it is among the most important. Dopamine may play a bigger role in the motivation to get a reward than the pleasure of the reward itself. *Wanting* more than *liking*. Genetically engineered mice unable to make dopamine will not seek out food, and will starve to death even when food is placed just inches from their mouth. Yet if food is put directly into their mouth, they will chew and eat the food, and seem to enjoy it.

Debates about differences between motivation and pleasure notwithstanding, dopamine is used to measure the addictive potential of any behavior or drug. The more dopamine a drug releases in the brain's reward pathway (a brain circuit that links the ventral tegmental area, the nucleus accumbens, and the prefrontal cortex), and the faster it releases dopamine, the more addictive the drug.

DOPAMINE REWARD PATHWAYS IN THE BRAIN

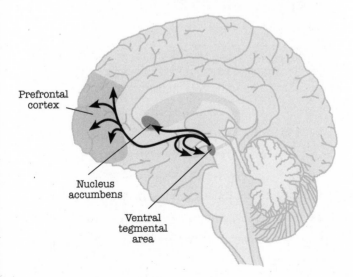

Prefrontal cortex

Nucleus accumbens

Ventral tegmental area

This is not to say that *high-dopamine* substances literally contain dopamine. Rather, they trigger the release of dopamine in our brain's reward pathway.

For a rat in a box, chocolate increases the basal output of dopamine in the brain by 55 percent, sex by 100 percent, nicotine by 150 percent, and cocaine by 225 percent. Amphetamine, the active ingredient in the street drugs "speed," "ice," and "shabu" as well as in medications like Adderall that are used to treat attention deficit disorder, increases the release of dopamine by 1,000 percent. By this accounting, one hit off a meth pipe is equal to ten orgasms.

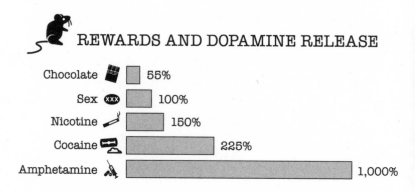

REWARDS AND DOPAMINE RELEASE

Chocolate 55%
Sex 100%
Nicotine 150%
Cocaine 225%
Amphetamine 1,000%

Pleasure and Pain Are Co-Located

In addition to the discovery of dopamine, neuroscientists have determined that pleasure and pain are processed in overlapping brain regions and work via an opponent-process mechanism. Another way to say this is that pleasure and pain work like a balance.

Imagine our brains contain a balance—a scale with a fulcrum

in the center. When nothing is on the balance, it's level with the ground. When we experience pleasure, dopamine is released in our reward pathway and the balance tips to the side of pleasure. The more our balance tips, and the faster it tips, the more pleasure we feel.

Pleasure   Pain

But here's the important thing about the balance: It wants to remain level, that is, in equilibrium. It does not want to be tipped for very long to one side or another. Hence, every time the balance tips toward pleasure, powerful self-regulating mechanisms kick into action to bring it level again. These self-regulating mechanisms do not require conscious thought or an act of will. They just happen, like a reflex.

I tend to imagine this self-regulating system as little gremlins hopping on the pain side of the balance to counteract the weight on the pleasure side. The gremlins represent the work of *homeostasis*: the tendency of any living system to maintain physiologic equilibrium.

Pleasure   Pain

Once the balance is level, it keeps going, tipping an equal and opposite amount to the side of pain.

Pleasure   Pain

In the 1970s, social scientists Richard Solomon and John Corbit called this reciprocal relationship between pleasure and pain the *opponent-process theory:* "Any prolonged or repeated departures from hedonic or affective neutrality . . . have a cost." That cost is an "after-reaction" that is opposite in value to the stimulus. Or as the old saying goes, *What goes up must come down.*

As it turns out, many physiologic processes in the body are governed by similar self-regulating systems. For example,

Johann Wolfgang von Goethe, Ewald Hering, and others have demonstrated how color perception is governed by an opponent-process system. Looking closely at one color for a sustained period spontaneously produces an image of its "opposing" color in the viewer's eye. Stare at a green image against a white background for a period of time, and then look away at a blank white page, and you will see how your brain creates a red afterimage. The perception of green gives way in succession to the perception of red. When green is turned on, red can't be, and vice versa.

Tolerance (Neuroadaptation)

We've all experienced craving in the aftermath of pleasure. Whether it's reaching for a second potato chip or clicking the link for another round of video games, it's natural to want to re-create those good feelings or try not to let them fade away. The simple solution is to keep eating, or playing, or watching, or reading. But there's a problem with that.

With repeated exposure to the same or similar pleasure stimulus, the initial deviation to the side of pleasure gets weaker and shorter and the after-response to the side of pain gets stronger and longer, a process scientists call *neuroadaptation*. That is, with repetition, our gremlins get bigger, faster, and more numerous, and we need more of our drug of choice to get the same effect.

Needing more of a substance to feel pleasure, or experiencing less pleasure at a given dose, is called *tolerance*. Tolerance is an important factor in the development of addiction.

Pleasure   Pain

For me, reading the *Twilight* saga for a second time was pleasurable but not as pleasurable as the first time. By the fourth time I read it (yes, I read the entire saga four times), my pleasure was significantly diminished. The rereading never quite measured up to that first go-round. Furthermore, each time I read it, I was left with a deeper sense of dissatisfaction in its aftermath and a stronger desire to regain the feeling I had while reading it the first time. As I became "tolerant" to *Twilight*, I was forced to seek out newer, more potent forms of the same drug to try to recapture that earlier feeling.

With prolonged, heavy drug use, the pleasure-pain balance eventually gets weighted to the side of pain. Our hedonic (pleasure) set point changes as our capacity to experience pleasure goes down and our vulnerability to pain goes up. You might think of this as the gremlins camped out on the pain side of the balance, inflatable mattresses and portable barbecues in tow.

Pleasure

Pain

I became acutely aware of this effect of high-dopamine addictive substances on the brain's reward pathway in the early 2000s, when I started seeing more patients coming in to clinic on high-dose, long-term opioid therapy (think OxyContin, Vicodin, morphine, fentanyl) for chronic pain. Despite prolonged and high-dose opioid medications, their pain had only gotten worse over time. Why? Because exposure to opioids had caused their brain to reset its pleasure-pain balance to the side of pain. Now their original pain was worse, and they had new pain in parts of their body that used to be pain free.

This phenomenon, widely observed and verified by animal studies, has come to be called *opioid-induced hyperalgesia*. *Algesia*, from the Greek word *algesis*, means sensitivity to pain. What's more, when these patients tapered off opioids, many of them experienced improvements in pain.

Neuroscientist Nora Volkow and colleagues have shown that heavy, prolonged consumption of high-dopamine substances eventually leads to a dopamine deficit state.

Volkow examined dopamine transmission in the brains of healthy controls compared to people addicted to a variety of drugs two weeks after they stopped using. The brain images are striking. In the brain pictures of healthy controls, a kidney-bean-shaped area of the brain associated with reward and motivation lights up bright red, indicating high levels of dopamine neurotransmitter activity. In the pictures of people with addiction who stopped using two weeks prior, the same kidney-bean-shaped region of the brain contains little or no red, indicating little or no dopamine transmission.

As Dr. Volkow and her colleagues wrote, "The decreases in DA D_2 receptors in the drug abusers, coupled to the decreases in DA release, would result in a decreased sensitivity of reward circuits to stimulation by natural rewards." Once this happens, nothing feels good anymore.

To put it another way, the players on Team Dopamine take their balls and their mitts and go home.

EFFECTS OF ADDICTION ON DOPAMINE RECEPTORS

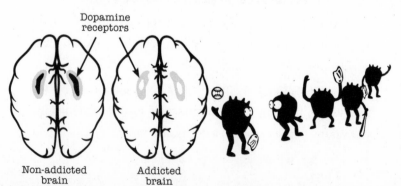

Dopamine receptors

Non-addicted brain

Addicted brain

In the approximately two years in which I compulsively consumed romance novels, I eventually reached a place where I could not find a book I enjoyed. It was as if I had burned out my novel-reading pleasure center, and no book could revive it.

The paradox is that hedonism, the pursuit of pleasure for its own sake, leads to *anhedonia*, which is the inability to enjoy pleasure of any kind. Reading had always been my primary source of pleasure and escape, so it was a shock and a grief when it stopped working. Even then it was hard to abandon.

My patients with addiction describe how they get to a point where their drug stops working for them. They get no high at all anymore. Yet if they don't take their drug, they feel miserable. The universal symptoms of withdrawal from any addictive substance are anxiety, irritability, insomnia, and dysphoria.

A pleasure-pain balance tilted to the side of pain is what drives people to relapse even after sustained periods of abstinence. When our balance is tilted to the pain side, we crave our drug just to feel normal (a level balance).

The neuroscientist George Koob calls this phenomenon "dysphoria driven relapse," wherein a return to using is driven not by the search for pleasure but by the desire to alleviate physical and psychological suffering of protracted withdrawal.

Here's the good news. If we wait long enough, our brains (usually) readapt to the absence of the drug and we reestablish our baseline homeostasis: a level balance. Once our balance is level, we are again able to take pleasure in everyday, simple rewards. Going for a walk. Watching the sun rise. Enjoying a meal with friends.

AHHH

Simple Pleasures

Pleasure   Pain

People, Places, and Things

The pleasure-pain balance is triggered not only by reexposure to the drug itself but also by exposure to cues associated with drug use. In Alcoholics Anonymous, the catchphrase to describe this phenomenon is *people, places, and things*. In the world of neuroscience, this is called *cue-dependent learning*, also known as classical (Pavlovian) conditioning.

Ivan Pavlov, who won the Nobel Prize in Physiology or Medicine in 1904, demonstrated that dogs reflexively salivate when presented with a slab of meat. When the presentation of meat is consistently paired with the sound of a buzzer, the dogs salivate when they hear the buzzer, even if no meat is immediately forthcoming. The interpretation is that the dogs have learned to associate the slab of meat, a natural reward, with the buzzer, a conditioned cue. What's happening in the brain?

By inserting a detection probe into a rat's brain, neuroscientists can demonstrate that dopamine is released in the brain in response to the conditioned cue (e.g., a buzzer, metronome, light) well before the reward itself is ingested (e.g., cocaine injection). The pre-reward dopamine spike in response to the conditioned cue explains the anticipatory pleasure we experience when we know good things are coming.

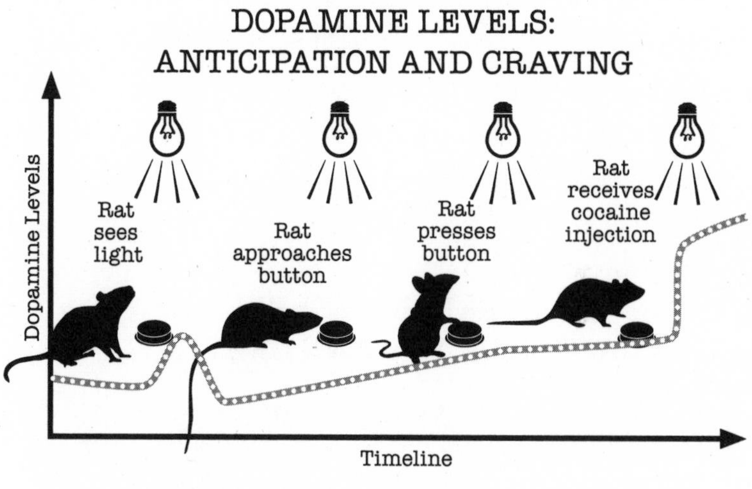

DOPAMINE LEVELS:
ANTICIPATION AND CRAVING

Dopamine Levels

Rat sees light

Rat approaches button

Rat presses button

Rat receives cocaine injection

Timeline

Right after the conditioned cue, brain dopamine firing decreases not just to baseline levels (the brain has a tonic level of dopamine firing even in the absence of rewards), but below baseline levels. This transient dopamine mini-deficit state is what motivates us to seek out our reward. Dopamine levels below baseline drive craving. Craving translates into purposeful activity to obtain the drug.

My colleague Rob Malenka, an esteemed neuroscientist, once said to me that "the measure of how addicted a laboratory animal is comes down to how hard that animal is willing to work to obtain its drug—by pressing a lever, navigating a maze, climbing up a chute." I've found the same to be true for humans. Not to mention that the entire cycle of anticipation and craving can occur outside the threshold of conscious awareness.

Once we get the anticipated reward, brain dopamine firing increases well above tonic baseline. But if the reward we anticipated doesn't materialize, dopamine levels fall well below baseline. Which is to say, if we get the expected reward, we get an even bigger spike. If we don't get the expected reward, we experience an even bigger plunge.

DOPAMINE LEVELS: ANTICIPATION AND CRAVING

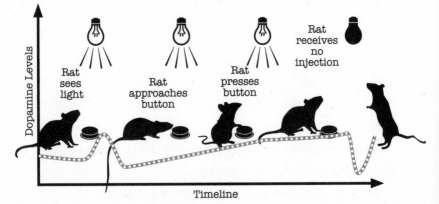

We've all experienced the letdown of unmet expectations. An expected reward that fails to materialize is worse than a reward that was never anticipated in the first place.

How does cue-induced craving translate to our pleasure-pain balance? The balance tips to the side of pleasure (a dopamine mini spike) in anticipation of future reward, immediately followed by a tip to the side of pain (a dopamine mini deficit) in the aftermath of the cue. The dopamine deficit is craving and drives drug-seeking behavior.

Over the past decade, significant advances have been made in understanding the biological cause of pathological gambling, leading to the reclassification of gambling disorders in the *Diagnostic and Statistical Manual of Mental Disorders* (5th edition) as addictive disorders.

Studies indicate that dopamine release as a result of gambling links to the unpredictability of the reward delivery, as much as to the final (often monetary) reward itself. The motivation to gamble is based largely on the inability to predict the reward occurrence, rather than on financial gain.

In a 2010 study, Jakob Linnet and his colleagues measured the dopamine release in people addicted to gambling and in healthy controls while winning and losing money. There were no distinct differences between the two groups when they won money; however, when compared to the control group, the pathological gamblers showed a marked increase in dopamine levels when they lost money. The amount of dopamine released in the reward pathway was at its highest when the probability of losing and winning was nearly identical (50 percent)—representing maximum uncertainty.

Gambling disorder highlights the subtle distinction between reward anticipation (dopamine release prior to reward) and reward response (dopamine release after or during reward). My patients with gambling addiction have told me that while playing, a part of them wants to lose. The more they lose, the stronger the urge to continue gambling, and the stronger the rush when they win—a phenomenon described as "loss chasing."

I suspect something similar is going on with social media apps, where the response of others is so capricious and unpredictable that the uncertainty of getting a "like" or some equivalent is as reinforcing as the "like" itself.

———

The brain encodes long-term memories of reward and their associated cues by changing the shape and size of dopamine-producing neurons. For example, the dendrites, the branches off the neuron, become longer and more numerous in response to high-dopamine rewards. This process is called *experience-dependent plasticity*. These brain changes can last a lifetime and persist long after the drug is no longer available.

Researchers explored the effects of cocaine exposure on rats by injecting them with the same amount of cocaine on successive days for a week and measuring how much they ran in response to each injection. A rat injected with cocaine will run across the cage instead of keeping to the periphery like normal rats do. The amount of running can be measured by using beams of light that project across the cage. The more times the rat breaks the beams of light, the more it's running.

The scientists found that with each successive day of cocaine exposure, the rats progressed from a lively jog on the first day, to an outright running frenzy on the last, showing a cumulative sensitization to the effects of cocaine.

Once the researchers stopped administering cocaine, the rats stopped running. One year later—a veritable lifetime for a rat—the scientists reinjected the rats with cocaine one time, and the rats were immediately running as they had on the final day of the original experiment.

When the scientists examined the rats' brains, they saw cocaine-induced changes in the rats' reward pathways consistent with persistent cocaine sensitization. These findings show that a drug like cocaine can alter the brain forever. Similar findings have been shown with other addictive substances, from alcohol to opioids to cannabis.

In my clinical work I see people who struggle with severe addiction slipping right back into compulsive use with a single exposure, even after years of abstinence. This may occur because of persistent sensitization to the drug of choice, the distant echoes of earlier drug use.

———

Learning also increases dopamine firing in the brain. Female rats housed for three months in a diverse, novel, and stimulating environment show a proliferation of dopamine-rich synapses in the brain's reward pathway compared to rats housed in standard laboratory cages. The brain changes that occur in response to a stimulating and novel environment are similar to those seen with high-dopamine (addictive) drugs.

But if the same rats are pretreated with a stimulant such as methamphetamine, a highly addictive drug, before entering the enriched environment, they fail to show the synaptic changes seen previously with exposure to the enriched environment. These findings suggest that methamphetamine limits a rat's ability to learn.

Here's some good news. My colleague Edie Sullivan, a world expert on alcohol's effects on the brain, has studied the process of recovery from addiction and discovered that although some brain changes due to addiction are irreversible, it is possible to detour around these damaged areas by creating new neural networks. This means that although the brain changes are permanent, we can find new synaptic pathways to create healthy behaviors.

Meanwhile, the future holds tantalizing possibilities for ways to reverse the scars of addiction. Vincent Pascoli and his colleagues injected rats with cocaine, which demonstrated the expected behavioral changes (frenzied running), then used optogenetics—a biological technique that involves the use of light to control neurons—to reverse the synaptic brain changes caused by cocaine. Maybe someday optogenetics will be possible on human brains.

The Balance Is Only a Metaphor

In real life, pleasure and pain are more complex than the workings of a balance.

What's pleasurable for one person may not be for another. Each person has their "drug of choice."

Pleasure and pain can occur simultaneously. For example, we can experience both pleasure and pain when eating spicy food.

Not everyone starts out with a level balance: Those with depression, anxiety, and chronic pain start with a balance tipped to the side of pain, which may explain why people with psychiatric disorders are more vulnerable to addiction.

Our sensory perception of pain (and pleasure) is heavily influenced by the meaning we ascribe to it.

Henry Knowles Beecher (1904–1976) served as a military doctor during World War II in North Africa, Italy, and France. He observed and reported on 225 soldiers with severe war-theater wounds.

Beecher was strict with his study inclusion criteria, surveying only those men who "had one of five kinds of severe wounds chosen as representative; extensive peripheral soft-tissue injury, compound fracture of a long bone, a penetrated head, a penetrated chest, or a penetrated abdomen . . . were clear mentally, and . . . were not in shock at the time of questioning."

Beecher made a remarkable discovery. Three-quarters of these badly injured soldiers reported little or no pain in the immediate aftermath of their wounds, despite life-threatening injuries.

He concluded that their physical pain was tempered by the emotional relief of escaping "from an exceedingly dangerous environment, one filled with fatigue, discomfort, anxiety, fear and real danger of death." Their pain, such as it was, gave them "a ticket to the safety of the hospital."

By contrast, a case report from the *British Medical Journal* published in 1995 details the case of a twenty-nine-year-old construction worker who walked into the emergency room after landing footfirst on a fifteen-centimeter nail, which was sticking up out of the top of his construction boot, having penetrated through leather, flesh, and bones. "The smallest movement of the nail was painful [and] he was sedated with fentanyl and midazolam," powerful opioids and sedatives.

But when the nail was pulled out from below and the boot removed, it became apparent that "the nail had penetrated between the toes: the foot was entirely uninjured."

Science teaches us that every pleasure exacts a price, and the pain that follows is longer lasting and more intense than the pleasure that gave rise to it.

With prolonged and repeated exposure to pleasurable stimuli, our capacity to tolerate pain decreases, and our threshold for experiencing pleasure increases.

By imprinting instant and permanent memory, we are unable to forget the lessons of pleasure and pain even when we want to: hippocampal tattoos to last a lifetime.

The phylogenetically uber-ancient neurological machinery for processing pleasure and pain has remained largely intact throughout evolution and across species. It is perfectly adapted for a world of scarcity. Without pleasure we wouldn't eat, drink, or reproduce. Without pain we wouldn't protect ourselves from injury and death. By raising our neural set point with repeated pleasures, we become endless strivers, never satisfied with what we have, always looking for more.

But herein lies the problem. Human beings, the ultimate seekers, have responded too well to the challenge of pursuing pleasure and avoiding pain. As a result, we've transformed the world from a place of scarcity to a place of overwhelming abundance.

Our brains are not evolved for this world of plenty. As Dr. Tom Finucane, who studies diabetes in the setting of chronic sedentary feeding, said, "We are cacti in the rain forest." And like cacti adapted to an arid climate, we are drowning in dopamine.

The net effect is that we now need more reward to feel pleasure, and less injury to feel pain. This recalibration is occurring not just at the level of the individual but also at the level of nations. Which invites the question: How do we survive and thrive in this new ecosystem? How do we raise our children? What new ways of thinking and acting will be required of us as denizens of the twenty-first century?

Who better to teach us how to avoid compulsive over-consumption than those most vulnerable to it: those struggling with addiction. Shunned for millennia across cultures as reprobates, parasites, pariahs, and purveyors of moral turpitude, people with addiction have evolved a wisdom perfectly suited to the age we live in now.

What follows are lessons of recovery for a reward-weary world.

PART II
Self-Binding

CHAPTER 4

Dopamine Fasting

I'm here because my parents made me come," Delilah said in that sullen voice that is the hallmark of the American teenager.

"Okay," I said. "Why do your parents want you to see me?"

"They think I'm smoking too much pot, but my problem is anxiety. I smoke because I'm anxious, and if you could do something about that, then I wouldn't need the weed."

I was gripped by a moment of overwhelming sadness. Not because I didn't know what to recommend, but because I was afraid she wouldn't take my advice.

"Okay, then let's start there," I said. "Tell me about your anxiety."

Long-limbed and graceful, she folded her legs underneath her.

"It started in junior high," she said, "and it's just gotten worse over the years. Anxiety is like the first thing I feel when I wake up in the morning. Hitting my wax pen is the only thing that gets me out of bed."

"Your wax pen?"

"Yeah, I vape now. I used to do blunts and bongs, Sativa in the daytime and Indica before bed. But now I'm into concentrates . . . wax, oil, budder, shatter, scissor, dust, QWISO. I mostly use a vape pen, but sometimes a Volcano. . . . I don't love edibles, but I'll use them in between or in an emergency when I can't smoke."

D *Stands for* Data

By prompting her to say more about her "wax pen," I was inviting Delilah to delve into the nitty-gritty details of her everyday use. My conversation with her was guided by a framework I've developed over the years for talking with patients about the problem of compulsive overconsumption.

This framework is easily remembered by the acronym DOPAMINE, which applies not just to conventional drugs like alcohol and nicotine but also to any high-dopamine substance or behavior we ingest too much of for too long, or simply wish we had a slightly less tortured relationship with. Although originally developed for my professional practice, I've also applied it to myself and my own maladaptive habits of consumption.

———

The *d* in DOPAMINE stands for *data*. I begin by gathering the simple facts of consumption. In Delilah's case, I explored what she was using, how much, and how often.

When it comes to cannabis, the dizzying list of products and delivery mechanisms that Delilah described is standard fare for my patients nowadays. Many of them have the equivalent of a PhD in Pot by the time they come to see me. In contrast to the 1960s, when recreational weekends-only use was normative, my patients start smoking the moment they wake up in the morning and keep going all day long until they go to bed again. This is concerning on many levels, not the least of which is that daily use has been linked to addiction.

For myself, I began to suspect I was teetering into the danger zone when reading romance novels started to take up hours a day and days at a time.

O *Stands for* Objectives

"What does smoking do for you?" I asked Delilah. "How does it help?"

"It's the only thing that works for my anxiety," she said. "Without it I'd be nonfunctional . . . I mean even more nonfunctional than I am now."

In asking Delilah to tell me how cannabis helped her, I was validating that it was doing something positive or she wouldn't be using it.

The *o* in DOPAMINE stands for *objectives* for using. Even seemingly irrational behavior is rooted in some personal logic. People use high-dopamine substances and behaviors for all kinds of reasons: to have fun, to fit in, to relieve boredom, to manage fear, anger, anxiety, insomnia, depression, inattention, pain, social phobia . . . the list goes on.

I used romance to escape what for me was a painful transition away from parenting young children to parenting teenagers, a job at which I felt much less skilled. I was also assuaging my grief at never having another baby, something I wanted and my husband did not, creating a tension in our marriage and in our sex life that hadn't existed before.

P *Stands for* Problems

"Any downsides from smoking? Unintended consequences?" I asked.

"The only bad thing about smoking," Delilah said, "is that my parents are always on my back. If they would just leave me alone, there wouldn't be any downsides."

I paused to notice the sun glinting on her hair. She was the picture of health despite the fact that she was ingesting more than a gram of cannabis a day. Youth, I thought, compensates for so much.

The *p* in DOPAMINE stands for *problems* related to use.

High-dopamine drugs always lead to problems. Health problems. Relationship problems. Moral problems. If not right away, then eventually. That Delilah could not see downsides—except the mounting conflict between her and her parents—is typical for teenagers . . . and not just teenagers. This disconnect occurs for a number of reasons.

First, most of us are unable to see the full extent of the consequences of our drug use while we're still using. High-dopamine substances and behaviors cloud our ability to accurately assess cause and effect.

As the neuroscientist Daniel Friedman, who studies the foraging practices of red harvester ants, once remarked to me, "The world is sensory rich and causal poor." That is to say, we know the doughnut tastes good in the moment, but we are less aware that eating a doughnut every day for a month adds five pounds to our waistline.

Second, young people, even heavy users, are more immune to the negative consequences of use. As one high school teacher remarked to me, "Some of my best students smoke pot every day."

As we age, however, the unintended consequences of chronic use multiply. Most of my patients who come in voluntarily for treatment are middle-aged. They seek me out because they've reached a tipping point where the downsides of their use outweigh the upsides. As they say in AA, "I'm sick and tired of being sick and tired." My teenage patients, by contrast, are neither sick nor tired.

Even so, getting teenagers to see some negative consequences of their use while they're still using, even if it's only that other people don't like it, can be a point of leverage for getting them to stop. And stopping, even just for a period of time, is essential for getting them to see true cause and effect.

A *Stands for* Abstinence

"I do have an idea about what might help you," I said to Delilah, "but it will require you to do something really hard."

"What's that?"

"I'd like you to try an experiment."

"An experiment?" She tilted her head to the side.

"I'd like you to stop using cannabis for a month."

Her face was impassive.

"Let me explain. First, treatments for anxiety are unlikely to work while you're smoking that much cannabis. Second, and more importantly, there's a distinct possibility that if you stop smoking for a whole month, your anxiety will get better all on its own. Of course, at first you'll feel worse due to withdrawal. But if you can get through the first two weeks, there's a good chance that in the second two weeks you'll start to feel better."

She remained quiet, so I continued. I explained to her that any drug that stimulates our reward pathway the way cannabis does has the potential to change our brain's baseline anxiety. What feels like cannabis treating anxiety may in fact be cannabis relieving withdrawal from our last dose. Cannabis becomes the cause of our anxiety rather than the cure. The only way to know for sure is to lay off for a month.

"Can I stop for a week?" she asked. "I've done that before."

"A week would be good, but in my experience, a month is usually the minimum amount of time it takes to reset the brain's reward pathway. If you don't feel better after four weeks of abstaining, that's also useful data. That means the cannabis isn't driving this, and we need to think about what else is. So what do you think? Do you think you would be able and willing to stop cannabis for a month?"

"Hmmm. . . . I don't think I'm ready to try quitting now, but maybe later. For sure I'm not going to be smoking like this forever."

"Do you still want to be using cannabis like this ten years from now?"

"No. No way. Definitely not." She shook her head vigorously.

"How about five years from now?"

"No, not in five years either."

"How about a year from now?"

Pause. Chuckle. "I guess you got me there, Doc. If I don't want to be using like this in a year, I might as well try to stop now."

She looked at me and smiled. "Okay, let's do this."

In asking Delilah to consider her current behavior in light of her future self, I hoped that quitting smoking would take on new urgency. It seemed to have worked.

———

The *a* in DOPAMINE stands for *abstinence*.

Abstinence is necessary to restore homeostasis, and with it our ability to get pleasure from less potent rewards, as well as see the true cause and effect between our substance use and the way we're feeling. To put it in terms of the pleasure-pain balance, fasting from dopamine allows sufficient time for the gremlins to hop off the balance and for the balance to go back to the level position.

The question is: How long do people need to abstain in order to experience the brain benefits of stopping?

Think back to the imaging study by neuroscientist Nora Volkow, showing that dopamine transmission is still below normal two weeks after quitting drugs. Her study is consistent with my clinical experience that two weeks of abstinence is not enough. At two weeks, patients are usually still experiencing withdrawal. They are still in a *dopamine deficit state*.

On the other hand, four weeks is often sufficient. Marc Schuckit and his colleagues studied a group of men who were drinking alcohol daily in large quantities and also met criteria for clinical depression, or what is called *major depressive disorder*.

Schuckit, a professor of experimental psychology at University of California, San Diego, is best known for demonstrating that biological sons of "alcoholics" have an increased genetic risk of developing an alcohol use disorder, compared to those without this genetic load. I had the pleasure of learning from Marc, a gifted teacher, at a series of conferences on addiction in the early 2000s.

The depressed men in Schuckit's study went into the hospital for four weeks, during which time they received no treatment for depression, other than stopping alcohol. After one month of not drinking, 80 percent no longer met criteria for clinical depression.

This finding implies that for the majority, clinical depression was the result of heavy drinking and not a co-occurring depressive disorder. Of course there are other explanations for these results: the therapeutic milieu of the hospital environment, spontaneous remission, the episodic nature of depression, which can come and go independent of external factors. But the robust findings are remarkable given that standard treatments for depression, whether medications or psychotherapy, have a 50 percent response rate.

Naturally I've seen patients who need less than four weeks to reset their reward pathway, and others who need far longer. Those who have been using more potent drugs in larger quantities for longer duration will typically need more time. Younger people recalibrate faster than older people, their brains being more plastic. Furthermore, physical withdrawal varies drug to drug. It can be minor for some drugs like video games but potentially life-threatening for others, like alcohol and benzodiazepines.

Which brings us to an important caveat: I never suggest a dopamine fast to individuals who might be at risk to suffer life-threatening withdrawal if they were to quit all of a sudden, as in cases of severe alcohol, benzodiazepine (Xanax, Valium, or Klonopin), or opioid dependence and withdrawal. For those patients, medically monitored tapering is necessary.

Sometimes, patients ask if they can swap one drug for another: cannabis for nicotine, video games for pornography. This is seldom an effective long-term strategy.

Any reward that is potent enough to overcome the gremlins and tip the balance toward pleasure can itself be addictive, thereby resulting in trading one addiction for another (cross addiction). Any reward that is not potent enough won't feel like a reward, which is why when we're consuming high-dopamine rewards, we lose the ability to take joy in ordinary pleasures.

Pleasure   Pain

A minority of patients (about 20 percent) don't feel better after the dopamine fast. That's important data too, because it tells me that the drug wasn't the main driver of the psychiatric symptom and that the patient likely has a co-occurring psychiatric disorder that will require its own treatment.

Even when the dopamine fast is beneficial, a co-occurring psychiatric disorder should be treated concurrently. Managing addiction without also addressing other psychiatric disorders typically leads to bad outcomes for both.

Nonetheless, to appreciate the relationship between the substance use and the psychiatric symptoms, I need to observe the patient for a sufficient period of time off high-dopamine rewards.

M *Stands for* Mindfulness

"I want you to be prepared," I said to Delilah, "for feeling worse before you feel better. By this I mean, when you first stop cannabis, your anxiety will get worse. But remember, this is not the anxiety you'll have to live with off cannabis. This is withdrawal-mediated anxiety. The longer you can go without using, the faster you'll get to that place where you're feeling better. Usually patients report a turning point at around two weeks."

"Okay. What am I supposed to do in the meantime? Do you have any pills you can give me?"

"There's nothing I can give you to take the pain away that's not also addictive. Since we don't want to trade one addiction for another, what I'm asking you to do is tolerate the pain."

Gulp.

"Yeah, I know. Hard. But it's also an opportunity. A chance for you to observe yourself as separate from your thoughts, emotions, and sensations, including pain. This practice is sometimes called *mindfulness.*"

The *m* of DOPAMINE stands for *mindfulness.*

Mindfulness is a term that is tossed around so often now, it has lost some of its meaning. Evolved from the Buddhist spiritual tradition of meditation, it has been adopted and adapted by the West as a wellness practice across many different disciplines. It has so fully penetrated Western consciousness that it's now routinely taught in American elementary schools. But what actually is mindfulness?

Mindfulness is simply the ability to observe what our brain is doing while it's doing it, without judgment. This is trickier than it sounds. The organ we use to observe the brain is the brain itself. Weird, right?

When I look at the Milky Way galaxy in the night sky, I'm always struck by how mysterious it is that we can be a part of something that looks so far away and separate. Practicing mindfulness is something like observing the Milky Way: It demands that we see our thoughts and emotions as separate from us and yet, simultaneously, a part of us.

Also, the brain can do some pretty weird things, some of which are embarrassing, hence the need for being *without judgment*. Reserving judgment is important to the practice of mindfulness because as soon as we start condemning what our brain is doing—*Ewww! Why would I be thinking about that? I'm a loser. I'm a freak*—we stop being able to observe. Staying in the observer position is essential to getting to know our brains and ourselves in a new way.

I remember standing in the kitchen in 2001 holding my newborn baby in my arms and experiencing an intrusive image of smashing her head against the refrigerator or the kitchen counter and watching it implode like a soft melon. The image was fleeting but vivid, and had I not been a regular practitioner of mindfulness, I would have done my best to ignore it.

Initially, I was horrified. As a psychiatrist I had treated mothers who, as a result of their mental illness, thought they had to kill their children to save the world. One of them actually did, an outcome I recall with sadness and regret to this day. So when I experienced an image of hurting my own child, I wondered if I was joining their ranks.

But remembering to observe without judgment, I followed the image and the feeling where they led and discovered that I didn't *want* to smash my baby's head; rather I had a great fear of doing so. The fear had manifested as the image.

Instead of condemning myself, I was able to have compassion for myself. I was grappling with the enormity of my responsibilities as a new mother, and what it meant to care for such a helpless creature, wholly dependent on me to protect her.

Mindfulness practices are especially important in the early days of abstinence. Many of us use high-dopamine substances and behaviors to distract ourselves from our own thoughts. When we first stop using dopamine to escape, those painful thoughts, emotions, and sensations come crashing down on us.

The trick is to stop running away from painful emotions, and instead allow ourselves to tolerate them. When we're able to do this, our experience takes on a new and unexpectedly rich texture. The pain is still there, but somehow transformed, seeming to encompass a vast landscape of communal suffering, rather than being wholly our own.

When I gave up my reading habit, I was gripped in the first several weeks by an existential terror. I lay on the couch in the evening, a time when I would normally reach for a book or some other method of distraction, with my hands folded over my stomach, trying to relax but instead feeling full of dread. I was astounded that such a seemingly small change in my daily routine could fill me with so much anxiety.

Then as the days passed and I continued the practice, I experienced a gradual relaxing of my mental boundaries and an opening up of my awareness. I began to see that I didn't need to continually distract myself from the present moment. That I could live in it and tolerate it, and maybe even something more.

I *Stands for* Insight

Delilah agreed to a month of abstinence. When she returned, her skin was glowing, the hunched shoulders were gone, and her sullen demeanor was replaced with a radiant smile. She strode into my office and took a chair.

"Well, I did it! And you're not going to believe this, Doc, but my anxiety is gone. Gone!"

"Tell me what happened."

"The first few days were bad. I felt blah. I threw up on the second day. Insane! I never throw up. I had this really sick feeling. That's when I realized I was withdrawing, and that motivated me to keep going with abstinence."

"Why would that motivate you?"

"Because it was the first piece of evidence I had that I was really addicted."

"So how did it go after that? How do you feel now?"

"Dude. So much better. Wow. Less anxiety. Definitely. That

word *anxiety* doesn't even come into my head. It used to rule my day. Clear-headed. I don't have to worry about my parents smelling it and getting mad. I'm not anxious at school anymore. The paranoia and suspiciousness . . . that's gone. I put so much time and mental effort into organizing my next high, rushing off to do it. It's such a relief not to have to do that anymore. I'm saving money. I've discovered events I enjoy more sober . . . like family events.

"Doctor, I'm telling you the truth, I did not see weed as a problem. I really didn't see it. But now that I've stopped smoking, I realize how much smoking was causing anxiety instead of curing it. I'd been smoking for five years without a break, and I didn't see what it was doing to me. I'm honestly kind of shocked."

———

The *i* of DOPAMINE stands for *insight*.

I have seen again and again in clinical care, and in my own life, how the simple exercise of abstaining from our drug of choice for at least four weeks gives clarifying insight into our behaviors. Insight that simply is not possible while we continue to use.

N *Stands for* Next Steps

As my visit with Delilah came to an end, I asked her about goals for the next month.

"So what do you think?" I said. "Do you want to continue to

abstain for the next month, or do you want to return to using?"

"Being sober," said Delilah, "I'm the best version of me."

I savored the moment.

"But," she said, "I still really like weed, and I miss the creative feeling it gives me, and the escape. I don't want to stop using. I'd like to go back to using, but not the way I was using before."

The *n* of DOPAMINE stands for *next steps*.

This is where I ask my patients what they want to do after their month of abstinence. The vast majority of my patients who are able to abstain for a month and experience the benefits of abstinence nonetheless want to go back to using their drug. But they want to use differently than they were using before. The overarching theme is that they want to use less.

An ongoing controversy in the field of addiction medicine is whether people who have been using drugs in an addictive way can return to moderate, nonrisky use. For decades the wisdom of Alcoholics Anonymous dictated that abstinence is the only option for people with addiction.

But emerging evidence suggests that some people who have met criteria for addiction in the past, especially those with less severe forms of addiction, can return to using their drug of choice in a controlled way. In my clinical experience, this has been true.

E *Stands for* Experiment

The *e* and final letter of DOPAMINE stands for *experiment*.

This is where patients go back out into the world armed with a new dopamine set point (a level pleasure-pain balance) and a plan for how to maintain it. Whether the goal is continued abstinence or moderation, like Delilah's, we strategize together for how to achieve it. Through a gradual process of trial and error, we figure out what works and what doesn't.

I would be remiss if I didn't point out that the goal of moderation, especially for people with severe addiction, can backfire, contributing to a precipitous escalation in use after a period of abstinence, sometimes referred to as the *abstinence violation effect*.

Rats who show a genetic propensity to become addicted will, after a two-to-four-week period of abstaining from alcohol, binge on alcohol as soon as they have access to it again, and continue to use heavily thereafter as if they had never abstained. A similar phenomenon has been observed in rats exposed to and hooked on high-calorie foods. The effect is attenuated in rats and mice less genetically predisposed to compulsive consumption.

What's not clear in animal studies is whether this binge-after-abstinence phenomenon is unique to drugs that are caloric, like food and alcohol, and not seen with noncaloric drugs like cocaine; or whether the real driver is the genetic predisposition of the rats themselves.

Even when moderation is achievable, many of my patients

report it's too exhausting to continue, and they ultimately opt for abstinence for the long haul.

But how about patients addicted to food? Or smartphones? Drugs that can't be stopped altogether?

The question of how to moderate is becoming an increasingly important one in modern-day life, because of the sheer ubiquity of high-dopamine goods, making us all more vulnerable to compulsive overconsumption, even when not meeting clinical criteria for addiction.

Further, as digital drugs like smartphones have become embedded into so many aspects of our lives, figuring out how to moderate their consumption, for ourselves and our children, has become a matter of urgency. To that end, I now introduce a taxonomy of self-binding strategies.

But before we talk about self-binding, let's review the steps of the dopamine fast, the ultimate goal of which is to restore a level balance (homeostasis) and renew our capacity to experience pleasure in many different forms.

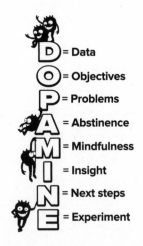

D = Data

O = Objectives

P = Problems

A = Abstinence

M = Mindfulness

I = Insight

N = Next steps

E = Experiment

CHAPTER 5

Space, Time, and Meaning

In the fall of 2017, after a year of abstaining from compulsive sexual behaviors, Jacob relapsed. He was sixty-five years old.

The trigger was a trip to Eastern Europe to see his family, complicated by his current wife and his children from his first marriage not getting along—the problem of money and who gets what, an old refrain.

Two weeks into his three-week trip, his children were angry because he had not given them the money they'd asked for. His wife was angry because he was even considering giving them money. He was afraid to disappoint anyone and hence threatened to disappoint them all.

He e-mailed me from overseas to let me know he was struggling. He hadn't relapsed yet but was close. I did some phone coaching and told him to come see me as soon as he got home. He came into the office a week after he returned, but by then it was too late.

"It is the TV in the hotel room that get me started craving

again," he said to me. "I want to watch the US Open. I lie there flipping through the channels, feeling depressed, thinking about my family, and my wife, and how everyone is angry at me. I see a naked woman on TV. Until I watch TV, I am pretty good. I am not getting urges. The biggest mistake is when I switch on the TV, I start thinking about returning to my old habits, and I can't stop the thoughts."

"Then what happened?"

"On Tuesday, I go home. I don't go to work. I stay home watching YouTube. I see body painting . . . people painting each other's naked bodies. A kind of art, I guess. On Wednesday, I cannot resist any longer. I go out and buy the parts to make my machine again."

"Your electrical stimulation machine?"

"Yes," he said sadly, only barely meeting my eyes. "The problem is when you start, you can be in ecstasy for a very long time. It's like being in a trance. And it's such a relief. I don't think about anything else. I go twenty hours without stopping. I go all day Wednesday and through the night. On Thursday morning, I throw the machine parts away in my garbage and go back to work. On Friday morning, I take them out of the garbage again and repair them and use all day. On Friday night, I call my sponsor, and go to a Sexaholics Anonymous meeting on Saturday. On Sunday, I take the parts out of the garbage and use again. And on Monday again. I want to stop but I can't. What should I do?"

"Pack up the machine and any spare parts," I told him, "and put it all in the garbage. Then take the garbage to the dump or somewhere else where it is impossible for you to retrieve it." He nodded understanding. "Then anytime you get the idea or urge or craving to use, drop to your knees and pray. Just pray. Ask God to help you, but do it from your knees. That's important."

I converged the mundane and the metaphysical. Nothing was too low or too high for my consideration. Telling him to pray was breaking unwritten rules, of course. Doctors don't talk about God. But I believe in believing, and my instincts told me this would resonate for Jacob, raised Roman Catholic.

Telling him to drop to his knees was also a way to insert some physicality into it, anything to break the mental compulsion that was compelling him to use. Or maybe I recognized some deeper need he had to act out his submission.

"After you've prayed," I said, "then get up and call your sponsor." He nodded again.

"Oh, and forgive yourself, Jacob. You're not a bad man. You've got problems, just like the rest of us."

Self-binding is the term to describe Jacob's act of throwing out his machine. It is the way we intentionally and willingly create barriers between ourselves and our drug of choice in order to mitigate compulsive overconsumption. Self-binding is not primarily a matter of will, although personal agency plays some part. Rather, self-binding openly recognizes the limitations of will.

The key to creating effective self-binding is first to acknowledge the loss of voluntariness we experience when under the spell of a powerful compulsion, and to bind ourselves while we still possess the capacity for voluntary choice.

If we wait until we feel the compulsion to use, the reflexive pull of seeking pleasure and/or avoiding pain is nearly impossible to resist. In the throes of desire, there's no deciding.

But by creating tangible barriers between ourselves and our drug of choice, we press the pause button between desire and action.

Further, self-binding has become a modern necessity. External rules and sanctions like taxes on cigarettes, age restrictions on alcohol, and laws prohibiting cocaine possession, although necessary, will never be sufficient in a world where access to an ever-growing variety of high-dopamine goods is practically infinite.

My patients have been telling me about their self-binding strategies for years. At some point I started writing them down. I repurpose strategies I learn from patients to advise other patients, as I did with Jacob when I told him to dispose of his machine in a remote dumpster that wouldn't allow him to retrieve it later.

I ask my patients, "What kinds of barriers can you put into place to make it harder for you to get easy access to your drug of choice?" I have even used self-binding in my own life to manage problems of compulsive overconsumption.

Self-binding can be organized into three broad categories: physical strategies (space), chronological strategies (time), and categorical strategies (meaning).

As you will see in what follows, self-binding is not fail-safe, particularly for those with severe addictions. It too can fall prey to self-deception, bad faith, and faulty science.

But it is a good and necessary place to start.

Physical Self-Binding

Of the many dangers that awaited Homer's Odysseus on his journey home from the Trojan War, the first was the Sirens, those half-woman, half-bird creatures whose enchanted song lured sailors to their death on the rocky cliffs of nearby islands.

The only way for a sailor to pass the Sirens unharmed was by not hearing them sing. Odysseus ordered his crew to put beeswax in their ears and tie him to the mast of the sailing ship, binding him even tighter if he begged to be unfastened or tried to break loose.

As this famous Greek myth illustrates, one form of self-binding is to create literal physical barriers and/or geographical distance between ourselves and our drug of choice. Here are some examples my patients have told me about: "I unplugged my TV and put it in my closet." "I banished my game console to the garage." "I don't use credit cards. Only cash." "I call hotels beforehand to ask them to remove the minibar." "I call hotels beforehand to ask them to remove the minibar *and* the television." "I put my iPad in a safety deposit box at Bank of America."

My patient Oscar, a rotund man in his late seventies with a

scholarly mind, a booming voice, and a penchant for talking in soliloquies, so much so that he made a muddle of group therapy and had to drop out, had a habit of drinking to excess while working in his study, tinkering in his garage, and puttering in his garden.

By trial and error he learned that to prevent this behavior, he had to remove all alcohol from his home. Any alcohol brought into the house needed to be locked up in a file cabinet for which only his wife had the key. Using this method, Oscar was able to successfully abstain from alcohol for years.

But I warned you that self-binding is no guarantee. Sometimes the barrier itself becomes an invitation to a challenge. Solving the puzzle of how to get our drug of choice becomes part of its appeal.

One day, Oscar's wife, on her way out of town, locked an expensive bottle of wine in a file cabinet and took the keys with her. The first evening she was away, Oscar got to thinking about the bottle of wine he knew was there. The thought intruded on his consciousness like a physical presence. It wasn't painful, just annoying. *If I just go take a peek and make sure it's all locked away, I'll stop thinking about it*, he told himself.

He walked to his wife's study and pulled on the drawer. To his surprise, the drawer opened half an inch, and he could see the bottle standing upright between the files. Not enough to get it out, but enough to see the cork, tantalizingly out of reach.

He stood staring into the darkened drawer for a full minute, contemplating the bottle. A part of him wanted to shut the drawer. Another part of him couldn't stop staring at it. Then something in his brain clicked and he decided—or maybe he stopped trying not to decide. He moved into action.

He hurried to the garage for his toolbox. Settling down to work, he used a wide range of tools to try to dismantle the lock and open the drawer. He worked with laser focus and determination. But he couldn't open the drawer. Every tool he tried failed to penetrate the lock.

Then the answer dawned on him like a knot suddenly coming loose under his fingers. *Of course. Why didn't I think of it before? It's so obvious.*

He sat up. No need to hurry now. His goal was in reach. He quietly packed up his tools save one, his long-stem pliers. He uncorked the bottle with the long-stem pliers, laid the cork and pliers gently on the table, and went to the kitchen to get the only remaining tool he would need: a long plastic straw.

Where Oscar's file cabinet failed, new devices like the kSafe kitchen safe might have done the trick. About the size of a bread box and made of impenetrable clear plastic, the kSafe holds everything from cookies to iPhones to opioid medication. A spin of the dial locks the safe on a timer. Once the timer has been set, there's no getting past the lock or penetrating the clear plastic material until the time is up.

Physical self-binding is now available from your local apothecary. Instead of locking our drugs away in a file cabinet, we have the option of imposing locks at the cellular level.

The medication naltrexone is used to treat alcohol and opioid addiction, and is being used for a variety of other addictions as well, from gambling to overeating to shopping. Naltrexone blocks the opioid receptor, which in turn diminishes the reinforcing effects of different types of rewarding behavior.

I've had patients report a near or complete cessation of alcohol craving with naltrexone. For patients who have struggled for decades with this problem, the ability to not drink at all, or to drink in moderation like "normal people," comes as a revelation.

Because naltrexone blocks our endogenous opioid system, people have reasonably wondered if it might induce depression. There's no reliable evidence of that, but I do occasionally see patients who report a flatlining of pleasure with naltrexone.

One patient said to me, "Naltrexone helps me not drink alcohol, but I don't enjoy bacon as much as I used to, or hot showers, and I can't get a runner's high." We worked around this by having him take naltrexone half an hour before entering a risky drinking situation, such as a happy hour. This naltrexone-as-needed approach allowed him to drink in moderation and also enjoy bacon again.

In the summer of 2014, one of my students and I traveled to China to interview people seeking treatment for heroin addiction at New Hospital, a voluntary, non-government-sponsored addiction treatment hospital in Beijing.

We talked to a thirty-eight-year-old man who described how

prior to coming to New Hospital for treatment, he had received the "addiction surgery." The addiction surgery involved insertion of a long-acting naltrexone implant to block the effects of heroin.

"In 2007," he said, "I went to Wuhan province for the surgery. My parents made me go, and they paid for it. I don't know for sure what the surgeons did, but I can tell you it didn't work. After the surgery, I kept shooting up heroin. I couldn't get the feeling anymore, but I did it anyway because shooting up was my habit. For the next six months I shot up every day with no feeling. I did not think about stopping because I still had money to buy it. After six months, the feeling came back. So I'm here now, hoping they'll have something new and better for me."

This anecdote illustrates that pharmacotherapy alone, without insight, understanding, and the will to change behavior, is unlikely to be successful.

Another medication that is used to treat alcohol addiction is disulfiram. Disulfiram interrupts alcohol metabolism, leading to the accumulation of acetaldehyde, which in turn causes a severe flushing reaction, nausea, vomiting, elevated blood pressure, and an overall feeling of malaise.

Taking disulfiram daily is an effective deterrent for those who are trying to abstain from alcohol, especially for people who wake up in the morning determined not to drink but by the evening have lost their resolve. It turns out that willpower is not an infinite human resource. It's more like exercising a muscle, and it can get tired the more we use it.

As one patient put it, "With disulfiram, I only need to decide once a day not to drink. I don't have to keep deciding all day long."

Some people, most commonly East Asians, have a genetic mutation that causes them to have a disulfiram-like reaction to alcohol without the drug. These individuals have historically had lower rates of alcohol addiction.

Of note, in recent decades, increased alcohol consumption in East Asian countries has led to higher rates of alcohol addiction even among this previously protected group. Scientists are now discovering that those with the mutation who drink anyway are at higher risk for alcohol-related cancers.

As with all forms of self-binding, disulfiram is fallible. My patient Arnold had been drinking heavily for decades, a problem that only got worse after he suffered a serious stroke and lost some of his frontal lobe function. His cardiologist told him he had to stop drinking or he would die. The stakes were high.

I prescribed disulfiram, and told Arnold the drug would make him sick if he drank while on it. In order to ensure Arnold took it, his wife administered it to him every morning and checked his mouth afterward to make sure he'd swallowed it.

One day while his wife was out, Arnold made his way over to the liquor store, got a fifth of whiskey, and drank it. When his wife came home and found him drunk, what puzzled her most was why the disulfiram hadn't made him sick. Arnold was intoxicated, but he wasn't ill.

A day later he confessed. For the preceding three days, he hadn't swallowed the pill. Instead, he'd wedged it in the gap left by a missing tooth.

Other modern forms of physical self-binding involve anatomical changes to our bodies; for example, weight-loss surgeries such as gastric banding, sleeve gastrectomy, and gastric bypass.

These surgeries effectively create a smaller stomach and/or bypass the part of the gut that absorbs calories. The gastric band puts a physical ring around the stomach, making it smaller without removing any part of the stomach or small intestine. The sleeve gastrectomy surgically removes part of the stomach to make it smaller. Gastric bypass surgery reroutes the small intestine around the stomach and duodenum, where nutrients are absorbed.

My patient Emily received gastric bypass surgery in 2014 and was thereby able to go from 250 pounds to 115 in the course of a year. No other interventions—and she had tried them all—had enabled her to lose weight. Emily is not alone.

Weight-loss surgeries are a proven effective intervention for obesity, especially when other remedies have failed. But they're not without unintended consequences.

One in four gastric bypass surgery recipients develops a new problem with alcohol addiction. In the wake of her surgery, Emily too became addicted to alcohol. The reasons are many.

Most people who are obese have an underlying food addiction, which is not adequately addressed with surgery alone. Few people who undergo these surgeries get the behavioral and psychological interventions they need to help them change their eating habits. Hence many of them resume eating in unhealthy ways, expand their now smaller stomachs, and end up with medical complications and the need for repeat surgeries. When food is no longer an option, many switch from food to another drug, like alcohol.

Further, the surgery alters how alcohol is metabolized, increasing the rate of absorption. The absence of a normal-size stomach means alcohol is absorbed into the bloodstream almost instantaneously and avoids the first-pass metabolism that usually occurs in the stomach. As a result, patients get intoxicated faster and stay intoxicated longer on less alcohol, akin to getting an alcohol IV.

We can and should celebrate a medical intervention that can improve the health of so many people. But the fact that we must resort to removing and reshaping internal organs to accommodate our food supply marks a turning point in the history of human consumption.

———

From lockboxes that limit our access, to medications that block our opioid receptors, to surgeries that shrink our stomachs, physical self-binding is everywhere in modern life, illustrating our growing need to put the brakes on dopamine.

As for me, when books were just one click away, I was prone to linger in fantasy longer than I wanted to, or than was good for me. I got rid of my Kindle and its easy access to a steady stream of downloadable erotica. As a result, I was better able to moderate my tendency to indulge in candy fiction. The simple act of having to go to the library or a bookstore created a useful barrier between me and my drug of choice.

Chronological Self-Binding

Another form of self-binding is the use of time limits and finish lines.

By restricting consumption to certain times of the day, week, month, or year, we narrow our window of consumption and thereby limit our use. For example, we can tell ourselves we'll consume only on holidays, only on weekends, never before Thursday, never before 5:00 p.m., and so on.

Sometimes, rather than time per se, we bind ourselves based on milestones or accomplishments. We'll wait till our birthday, or as soon as we complete an assignment, or after we get our degree, or once we get the promotion. When the clock has run down, or we've crossed a self-designated finish line, the drug is our reward.

Neuroscientists S. H. Ahmed and George Koob have shown that rats given unlimited access to cocaine for six hours per day gradually increase their lever-pressing over time to the point of physical exhaustion and even death. Increased self-administration under extended access conditions (six hours) has also been observed with methamphetamine, nicotine, heroin, and alcohol.

However, rats who have access to cocaine for only one hour per day use steady amounts of cocaine over many consecutive days. That is, they don't press the lever for more drug per unit time with each consecutive day.

This study suggests that by restricting drug consumption to a narrow window of time, we may be able to moderate our use and avoid the compulsive and escalating consumption that comes with unlimited access.

———

Just tracking how much time we spend consuming, for example, by clocking our smartphone use, is one way to become aware of and thereby mitigate consumption. When we make conscious use of objective facts like how much time we're using, we are less able to deny them, and therefore in a better position to take action.

However, this can get very tricky very fast. Time has a funny way of getting away from us when we're chasing dopamine.

One patient told me that when he was using methamphetamine, he convinced himself that time didn't count. He felt as though he could stitch it back together later without anyone realizing a piece had gone missing. I imagined him floating in the night sky, big as a constellation, sewing together a rent in the universe.

High-dopamine goods mess with our ability to delay gratification, a phenomenon called *delay discounting*.

Delay discounting refers to the fact that the value of a reward goes down the longer we have to wait for it. Most of us would rather get twenty dollars today than a year from now.

Our tendency to overvalue short-term rewards over longer-term ones can be influenced by many factors. One of those factors is consumption of addictive drugs and behaviors.

Behavioral economist Anne Line Bretteville-Jensen and her colleagues investigated the discounting in active heroin and amphetamine users compared with ex-users and with matched controls (individuals matched for gender, age, education level, etc.). The investigators asked the participants to imagine they had a winning lottery ticket worth 100,000 Norwegian kroner (NOK), approximately 14,600 US dollars.

They then asked participants if they would rather have less money right now (less than 100,000 NOK) or the full amount a week from now. Of active drug users, 20 percent said they wanted the money right now and would be willing to take less to get it. Only 4 percent of former users and 2 percent of matched controls would have accepted that loss.

Cigarette smokers are more likely than matched controls to discount monetary rewards (that is, they value them less if they have to wait longer for them). The more they smoke, and the more nicotine they consume, the more they discount future rewards. These findings hold true for both hypothetical money and real money.

Addictions researcher Warren K. Bickel and his colleagues asked people addicted to opioids and healthy controls to complete a story that started with the line: "After awakening, Bill began to think about his future. In general, he expected to . . ."

Opioid-addicted study participants referred to a future that was on average nine days long. Healthy controls referred to a future that was on average 4.7 years long. This striking difference illustrates how "temporal horizons" shrink when we're under the sway of an addictive drug.

Conversely, when I ask my patients what was the deciding moment for them to try to get into recovery, they'll say something that expresses a long view of time. As one patient told me who'd been snorting heroin for the past year, "I suddenly realized I'd been using heroin for a year, and I thought to myself, if I don't stop now, I may be doing this for the rest of my life."

Reflecting on the trajectory of his whole life, rather than just the present moment, allowed this young man to take a more accurate inventory of his day-to-day behaviors. The same was true of Delilah, who was willing to abstain from cannabis for four weeks only after imagining herself still smoking ten years hence.

In today's dopamine-rich ecosystem, we've all become primed for immediate gratification. We want to buy something, and the next day it shows up on our doorstep. We want to know something, and the next second the answer appears on our screen. Are we losing the knack of puzzling things out, or being frustrated while we search for the answer, or having to wait for the things we want?

The neuroscientist Samuel McClure and his colleagues examined what parts of the brain are involved in choosing immediate versus delayed rewards. They found that when participants chose immediate rewards, emotion- and reward-processing parts of the brain lit up. When participants delayed their reward, the prefrontal cortex—the part of the brain involved in planning and abstract thinking—became active.

The implication here is that we are all now vulnerable to prefrontal cortical atrophy as our reward pathway has become the dominant driver of our lives.

Ingestion of high-dopamine goods is not the only variable that influences delay discounting.

For example, those who grow up in resource-poor environments and are primed with mortality cues are more likely to value immediate rewards over delayed rewards compared to those who are similarly primed and grow up in resource-rich environments. Young Brazilians living in favelas (slums) discount future rewards more than age-matched university students.

Is it any wonder that poverty is a risk factor for addiction, especially in a world of easy access to cheap dopamine?

———

Another variable contributing to the problem of compulsive overconsumption is the growing amount of leisure time we have today, and with it the ensuing boredom.

The mechanization of agriculture, manufacturing, domestic chores, and many other previously time-consuming, labor-intensive jobs has reduced the number of hours per day people spend working, leaving more time for leisure.

A typical day for the average laborer in the United States

just before the Civil War (1861–1865), whether in agriculture or industry, consisted of working ten to twelve hours a day, six and a half days per week, fifty-one weeks per year, with no more than two hours a day spent on leisure activity. Some workers, often immigrant women, worked thirteen hours a day, six days a week. Others labored in slavery.

By contrast, the amount of leisure time in the United States today increased by 5.1 hours per week between 1965 and 2003, an additional 270 leisure hours per year. By 2040, the number of leisure hours in a typical day in the United States is projected to be 7.2 hours, with just 3.8 hours of daily work. The numbers for other high-income countries are similar.

Leisure time in the United States differs by education and socioeconomic status, but not in the way you might think.

In 1965, both the less educated and more educated in the United States enjoyed about the same amount of leisure time. Today, adults living in the US without a high school diploma have 42 percent more leisure time than adults with a bachelor's degree or higher, with the biggest differences in leisure time occurring during weekday hours. This is due in large part to underemployment among those without a college degree.

Dopamine consumption is not just a way to fill the hours not spent working. It has also become a reason why people are not participating in the workforce.

Economist Mark Aguiar and his colleagues wrote in an article aptly titled "Leisure Luxuries and the Labor Supply of Young Men," "Younger men, ages 21 to 30, exhibited a larger decline in work hours over the last fifteen years than older men or women. Since 2004, time-use data show that younger men distinctly shifted their leisure to video gaming and other recreational computer activities."

Writer Eric J. Iannelli briefly alluded to his own history of addiction as follows:

> Years ago, in what now seems like another life, a friend said to me, "Your entire existence can be reduced to a three-part cycle. One: Get fucked up. Two: Fuck up. Three: Damage control." We hadn't known each other very long, probably two months at most, and yet he had already witnessed enough of my regular blackout drinking, just one of the more obvious manifestations of addiction's self-perpetuating vortex, to have got my number. With a wry smile, he went on to hypothesize more generally—and, I suspect, only half-jokingly—that addicts are bored or frustrated problem-solvers who instinctively contrive Houdini-like situations from which to disentangle themselves when no other challenge happens to present itself. The drug becomes the reward when they succeed and the consolation prize when they fail.

———

When I first met Muhammad, he was a river of words. His tongue could barely keep up with his brain, which was teeming with ideas.

"I think I may have a little addiction problem," he said. I liked him immediately.

In flawless English with a slight Middle Eastern accent, he told me his story.

He came to the United States from the Middle East in 2007 to study undergraduate math and engineering. In his home country, drug use of any kind risked harsh punishment.

After arriving in the United States, it was liberating for him to be able to use drugs recreationally without fear. To begin, he restricted drug and alcohol use to the weekends, but within the year, he was smoking cannabis daily and could see that his grades and his friendships suffered as a result.

He told himself, *I'm not going to smoke again until I complete my undergraduate degree, get accepted to a master's program, and get funded for a PhD.*

True to his promise, he did not smoke again until he completed a Stanford master's program in mechanical engineering and got funding for a PhD. When he resumed smoking, he pledged to limit himself to weekends only.

A year into his PhD, he was smoking every day, and by the end of his second year, he set new rules for himself: *ten-milligram joints while working, thirty-milligram joints when not working, and three-hundred-milligram joints only on special occasions . . . to get really fucked up.*

Muhammad failed his qualifying exam, the culmination of his PhD studies. He took it a second time and failed again. He was about to be terminated from the program but managed to convince his professors to give him one last try.

In the spring of 2015, Muhammad committed to abstaining until he passed his qualifying exam, however long it took. For the next year, he abstained from cannabis and worked harder than ever before. His final report was over 100 pages long.

"It was," he told me, "one of the most positive and productive years of my life."

That year he passed his qualifying exams, and the night after his exam, a friend brought cannabis over to help him celebrate. At first, Muhammad declined. But his friend said, "There's no way someone as smart as you can be addicted."

Just this once, Muhammad told himself, *and then not again till graduation.*

By Monday, *not again till graduation* became *no marijuana on days that I have classes*, which became *no marijuana on days that I have hard classes*, which became *no marijuana on days that I have exams*, which became *no marijuana before nine a.m.*

Muhammad *was* smart. So why couldn't he figure out that every time he smoked, he wouldn't be able to stick to his self-imposed time limits?

Because once he started using cannabis, he wasn't governed by reason; he was governed by the pleasure-pain balance. Even one joint created a state of wanting not easily influenced by logic. Under the influence, he could no longer objectively evaluate the immediate rewards of smoking against their long-term counterparts. Delay discounting ruled his world.

In Muhammad's case, chronological self-binding went only so far, and cannabis in moderation was unlikely ever to be an option. He would have to, and eventually did, find another way.

Categorical Self-Binding

Jacob came to see me again a week after his relapse. He hadn't used for the entire week. He put his machine in a garbage can that he knew was being carted away the same day. He also put his laptop and tablet away. He went to church for the first time in years and prayed for his family.

"Not thinking about myself and my problems was a good change. I also stop shaming myself. Mine is a sad story, but I can do something about it."

He paused. "But I'm not feeling good," he said. "I see you on a Monday, and by Friday I think about killing myself, but I know I won't do it."

"It's the comedown from using," I said. "Let your feelings crest over you like a wave. Be patient, and with time, you will feel better."

In the weeks and months that followed, Jacob was able to maintain abstinence by limiting not just access to pornography, chat rooms, and TENS units, but also to "lust in any form."

He stopped watching television, movies, YouTube, women's volleyball competitions—pretty much anything that presented for him a sexually provocative image. He skipped over certain types of news articles; for example, articles about Stormy Daniels, the stripper who allegedly had an affair with Donald Trump. He put his shorts on before shaving in front of the mirror in the mornings. To see his own nakedness was itself a trigger.

"I played with my own body for a long time. I can't do that

anymore," he said. "I must avoid anything that might enter-tain my addict mind."

———

Categorical self-binding limits consumption by sorting dopa-mine into different categories: those subtypes we allow our-selves to consume, and those we do not.

This method helps us to avoid not only our drug of choice but also the triggers that lead to craving for our drug. This strategy is especially useful for substances we can't eliminate altogether but that we're trying to consume in a healthier way, like food, sex, and smartphones.

My patient Mitch was addicted to sports betting. He had lost a million dollars gambling by the time he was forty. Par-ticipating in Gamblers Anonymous was an important part of his recovery. Through his involvement in Gamblers Anony-mous, he learned that it wasn't just betting on sports he had to avoid. He also had to abstain from watching sports on TV, reading the sports page in the newspaper, surfing sports-related Internet sites, and listening to sports radio. He called all the casinos in his area and had himself put on the "no-admit" list. By avoiding substances and behaviors beyond his drug of choice, Mitch was able to use categorical binding to mitigate the risk of relapse to sports betting.

There's something tragic and touching about having to ban yourself.

As for Jacob, hiding the naked body, his and others', was an important part of his recovery. Concealing the body as a way to minimize the risk of engaging in forbidden sexual con-course has long been a part of many cultural traditions, continuing to the present day. The Quran says of female modesty: "And tell the believing women to cast down their glances and guard their private parts and not expose their adornment . . . and to wrap [a portion of] their headcovers over their chests and not expose their adornment."

The Church of Jesus Christ of Latter-day Saints (LDS Church) has issued official statements on modest dress for its members, such as discouraging "short shorts and short skirts, shirts that do not cover the stomach, and clothing that does not cover the shoulders or is low-cut in the front or the back."

———

Categorical self-binding fails when we inadvertently include a trigger in our list of acceptable activities. We can correct mistakes like these with a mental sifting process based on experience. But what about when the category itself changes?

The well-worn American tradition of dieting—vegetarian, vegan, raw vegan, gluten-free, Atkins, Zone, ketogenic, Paleolithic, grapefruit—is one example of categorical self-binding. We pursue these diets for varied reasons: medical, ethical, religious. But whatever the reason, the net effect is to decrease access to large food categories, which in turn limits consumption.

But diets as a form of categorical self-binding are threatened when the category changes over time as a result of market forces.

More than 15 percent of North American households use gluten-free products. Some people are gluten-free because they have Celiac disease, an autoimmune disease wherein the ingestion of gluten leads to damage in the small intestine. But growing numbers of people are gluten-free because it helps them limit consumption of high-calorie, low-nutrition carbohydrates. The problem?

Around 3,000 new gluten-free snack products were introduced in the US from 2008 to 2010, and bakery products are the single highest-grossing packaged-good category in the gluten-free market today. In 2020, the gluten-free products' value in the US alone was estimated at $10.3 billion.

A gluten-free diet, which previously had effectively limited consumption of high-calorie processed foods such as cakes, cookies, crackers, cereal, pastas, and pizzas, now no longer does. For those who were using the gluten-free diet to avoid gluten, this might be good news. But for those who were benefiting from gluten-free as a category to limit consumption of bread, cakes, and cookies, the category no longer serves.

The evolution of the gluten-free diet illustrates how attempts to control consumption are swiftly countered by modern market forces, just one more example of the challenges inherent in our dopamine economy.

There are many other modern examples of previously taboo drugs being transformed into socially acceptable commodities, often in the guise of medicines. Cigarettes became vape pens and ZYN pouches. Heroin became OxyContin. Cannabis became "medical marijuana." No sooner have we committed to abstinence than our old drug reappears as a nicely packaged, affordable new product saying, *Hey! This is okay. I'm good for you now.*

Deifying the demonized is another form of categorical self-binding.

Since prehistoric times, humans have elevated mind-altering drugs to sacred categories to be used during religious ceremonies, rites of passage, or as medicines. In this context, only priests, shamans, or other designates who have received special training or have been invested with special authority are allowed to administer these drugs.

For more than 7,000 years, hallucinogens, also known as psychedelics (magic mushrooms, ayahuasca, peyote), have had sacramental uses across diverse cultures. When hallucinogens became popular and widely available as recreational drugs in the counterculture movement of the 1960s, however, harms multiplied, leading to LSD being made illegal in most parts of the world.

Today, there is a movement to bring hallucinogens and other psychedelics back into use, but only in the pseudo-sacred context of psychedelic-assisted psychotherapy. Specially trained psychiatrists and psychologists are now administering hallucinogens and other potent psychotropic agents (psilocybin, ketamine, ecstasy) as mental health remedies. Administering limited doses (one to three) of psychedelics interspersed with multiple sessions of talk therapy over many weeks has become the modern equivalent of shamanism.

The hope is that by limiting access to these drugs, and by making psychiatrists gatekeepers, the mystical properties of these chemicals—a sense of oneness, transcendence of time, positive mood, and reverence—can be leveraged without leading to misuse, overuse, and addictive use.

———

Some people need neither shaman nor psychiatrist to imbue their drug of choice with the sacred. In a now famous Stanford marshmallow experiment, at least one child in the experiment managed the sacred entirely on their own.

The Stanford marshmallow experiment was a series of studies led by psychologist Walter Mischel in the late 1960s at Stanford University to study delayed gratification.

Children between the ages of three and six were offered a choice between one small reward provided immediately (a marshmallow) or two small rewards (two marshmallows) if the child could wait for approximately fifteen minutes without eating the first marshmallow.

During that time, the researcher left the room and then returned. The marshmallow was placed on a plate on a table in a room that was otherwise empty of distractions: no toys, no other children. The purpose of the study was to determine when delayed gratification develops in children. Subsequent studies examined what kinds of real-life outcomes are associated with the ability, or lack thereof, to delay gratification.

The researchers discovered that of approximately one hundred children, one-third made it long enough to get the second marshmallow. Age was a major determinant: the older the child, the more able to delay. In follow-up studies, children who were able to wait for the second marshmallow tended to have better SAT scores and better educational attainment, and were overall cognitively and socially better-adjusted adolescents.

One detail of the experiment that is less well known is what the children did during those fifteen minutes of struggling not to eat the first marshmallow.

The researchers' observations reveal a literal embodiment of self-binding: The children "cover their eyes with their hands or turn around so that they can't see the tray . . . start kicking the desk, or tug on their pigtails, or stroke the marshmallow as if it were a tiny stuffed animal."

Covering eyes and turning away is reminiscent of physical self-binding. Tugging on pigtails suggests using physical pain as a distraction . . . something I'll talk about later at length. But what of stroking the marshmallow? This child, instead of turning away from the desired object, made it a pet, far too precious to eat, or at least to eat impulsively.

My patient Jasmine came to me seeking help for excessive alcohol consumption, up to ten beers every day. As part of the treatment, I advised her to remove all alcohol from her home as a self-binding strategy. She mostly took my advice, with a twist.

She removed all alcohol save one beer, which she left in her refrigerator. She called it her "totemic beer," which she regarded as the symbol of her choice not to drink, a representation of her will and autonomy. She told herself that she only needed to focus on not drinking that one beer rather than the more daunting task of not drinking any beer from the vast quantity available in the world.

This metacognitive sleight of hand, transforming an object of temptation into a symbol of restraint, helped Jasmine abstain.

———

Half a year into his second attempt at recovery, I met Jacob in the waiting room. It had been several months since I had seen him.

As soon as I laid eyes on him, I knew he was doing well. It was the way his clothes fit him, the way they hugged his body. But it wasn't just his clothes. His skin fit him too, the way it does when a person feels connected to themselves and the world.

Not that you'll find that in any psychiatry textbook. It's just something I've noticed after decades seeing patients: When people get better, everything holds together and has a rightness. Jacob had a rightness to him that day.

"My wife is back in my life," he said once we were in my office. "We still living separately, but I go to Seattle to see her and we spend two wonderful days. We are going to spend the Christmas together."

"I'm glad, Jacob."

"I am free of my obsession. I am not compelled to behave in

a certain way. I am free to make decisions again about what I will do. I've got almost six months since my relapse. If I just keep doing what I am doing, I think I'm going to be okay. Better than okay."

He looked at me and smiled. I smiled back.

───────

The extraordinary lengths to which Jacob went to avoid anything likely to incite sexual desire seem downright medieval to our modern sensibilities, just one step removed from a hair shirt.

Yet far from feeling constrained by his new way of living, he felt liberated. Released from the grips of compulsive overconsumption, he was again able to interact with other people and the world with joy, curiosity, and spontaneity. He felt a certain dignity.

As Immanuel Kant wrote in *The Metaphysics of Morals*, "When we realize that we are capable of this inner legislation, the (natural) man feels himself compelled to reverence for the moral man in his own person."

Binding ourselves is a way to be free.

CHAPTER 6

A Broken Balance?

I'm hoping," Chris said, sitting in my office, adjusting his backpack, pushing back the hair that had fallen into his eyes, jangling his knee (I would learn over the ensuing years that he was always in motion), "that you will continue my buprenorphine. It's been helpful. Actually, that's an understatement. I'm not sure I'd be alive without it, and I need to find someone who can prescribe it for me."

Buprenorphine is a semisynthetic opioid derived from thebaine, distilled from the opium poppy. Like other opioids, buprenorphine binds to the μ-opioid receptor, providing immediate relief from pain and opioid craving. In the simplest terms, it works by bringing the pleasure-pain balance back to a level position, so that someone like Chris can stop battling craving and get back to living his life. The evidence is robust that buprenorphine decreases illicit opioid use, reduces the risk of overdose, and improves quality of life.

But there's no glossing over the fact that buprenorphine is an opioid that can be misused, diverted, and sold on the street. For people who aren't dependent on stronger opioids, buprenorphine can create a euphoric high. People on buprenorphine experience opioid withdrawal and craving when they stop or decrease the dose. In fact, I've had some patients tell me that the withdrawal from buprenorphine is far worse than anything they experienced with heroin or OxyContin.

"Why don't you tell me your story," I said to Chris, "and then I'll let you know what I think."

Chris arrived at Stanford in 2003. His stepfather drove him up from Arkansas in an old borrowed Chevy Suburban. The SUV, packed full of Chris's belongings, stood out among the shiny new BMWs and Lexuses crowding the entrance to student housing.

Chris didn't waste time. He organized his dorm room with meticulous precision, starting with his CD collection, which he arranged in alphabetical order. He studied the course catalogue and settled on creative writing, Greek philosophy, and Myth and Modernity in German Culture. He dreamed of becoming a composer, a film director, an author. His plans, like those of his fellow students, were grand. This would be his illustrious Stanford beginning.

Once classes began, Chris did well in all the expected ways. He studied hard. He got excellent grades. But on another level, he was not thriving: He attended his classes alone, studied in his room or the library alone, played the piano in the common room of his dorm alone. That favorite campus buzz-word, *community*, eluded him.

Most of us looking back on our early college days will remember struggling to find our people. Chris struggled more. It's hard to say, even now, exactly why. He's a good-looking young man. Thoughtful. Affable. Eager to please. Perhaps it had something to do with being that poor kid from Arkansas.

His solitary campus existence continued into his sophomore year until he met a girl at his part-time campus job. His chiseled features, soft brown hair, and wiry, muscular build had always attracted attention. He and the girl, a fellow undergraduate, kissed, and Chris fell instantly in love. When she told him she had a boyfriend, he decided it didn't matter. He wanted to be with her, and repeatedly sought her out. When he didn't give up, she accused him of stalking her and reported him to their mutual boss. As a result, he lost his job and was reprimanded by school administration. Without a job or a girlfriend, he decided there was only one solution: He would kill himself.

Chris wrote a parting e-mail to his mother: "Ma, I wore clean underwear." He borrowed a knife, took his CD player and a carefully selected CD, and made his way to Roble Field. It was dusk, and his plan was to swallow a bottle of pills, cut his wrists, and time his death with the setting sun.

Music was important to Chris, and he chose his final song with care: "PDA" by Interpol, a New York indie post-punk revival band. "PDA" is rhythmic and pounding. The lyrics are hard to make out. The last stanza goes like this: "Sleep tonight, sleep tonight, sleep tonight, sleep tonight. Something to say, something to do, nothing to say, there's nothing to do." Chris waited till the very end of the song, then pulled the sharp edge of the knife across each wrist.

Trying to kill yourself by slitting your wrists in an open field turns out not to be a very effective strategy. Half an hour later, the blood on his wrists had congealed, and he was sitting in the dark, watching people walk by. He went back to his dorm room, made himself vomit up the pills, and called 911. The paramedics came and took him to Stanford Hospital, where he got admitted to the psych ward.

His stepfather was the first to visit him. His mother planned to come too but was unable to board the plane. She had a long-standing fear of flying. His biological father, whom he saw only several times per year, also showed up. His father looked stricken when he saw the red, raised incisions on Christopher's wrists.

Chris stayed on the psychiatric ward for a total of two weeks. During that time, he mostly felt relieved to be in a contained, controlled, and predictable environment.

A representative from Stanford University came to visit him on the unit and informed him that, under the circumstances, he would be forced to take a medical leave from Stanford, until he recovered sufficiently to be able to return, at the determination and discretion of the university.

Chris went back to Arkansas to live with his mother and stepfather. He got a job waiting tables. He discovered drugs.

In the fall of 2007, Chris returned to Stanford. Before he

could enroll for the fall quarter, he needed to meet with the head of student mental health and his resident dean to update them on his progress and present a convincing argument for reenrolling.

The day before his meeting, he stayed with a girl he had known at Stanford. He hadn't known her well, but she was "troubled too," so Chris felt more comfortable asking if he could crash at her place for a night or two while he got himself squared away with the university.

The night before his interview, Chris stayed up "doing coke" and reading Freud's *Civilization and Its Discontents*. By morning he concluded he was too messed up to meet with a bunch of college administrators. He flew home the same day.

Chris spent the next year shoveling dirt, spreading mulch, and mowing lawns in 100+ degree weather for the University of Arkansas. He liked the physicality of it, the way that moving his body distracted him from his thoughts. He got promoted to arborist, which mostly involved shoving tree trunks and branches into a wood chipper.

When he wasn't working, he was composing music, score after score, while smoking cannabis, which had become indispensable to him.

Chris returned to Stanford again the next fall. No in-person meeting was required this time. Chris showed up to his dorm Jack Reacher–style, nothing but a toothbrush in his pocket and a laptop in his hand. He slept on his mattress in his clothes, no sheets.

He willed himself to be structured, something he recognized he would need to be successful. As part of his new mindset, he changed his major. He would study chemistry now.

He also vowed to quit smoking cannabis, but his resolve lasted only three days before he was back to smoking daily, hiding out in his room, trying to time it for when his roommate, whom he remembered merely as "some Indian guy," wasn't around.

At midterm time, Chris reasoned that since he'd spent most of his study time high, he should be high for his midterms. Something about "state-dependent learning" that he'd read about in his psychology class. He made it to the second question before realizing he didn't know the material and was unable to complete the exam. He stood up and walked out, throwing his test in the garbage on the way.

He was on a plane home the next day.

Leaving Stanford the third time felt different for Chris. It was tinged with hopelessness. When he got home, he had no ambition at all, not even to continue composing music. He began drinking heavily, in addition to smoking cannabis. Then he tried opioids for the first time, which was easy to do in Arkansas in 2009, when opioid manufacturers and distributors were pumping millions of opioid pain pills into the state. In that same year, doctors in Arkansas wrote 116 opioid prescriptions per 100 persons living in Arkansas.

While taking opioids, everything Chris thought he had been searching for suddenly seemed just within reach. Yes, he felt euphoric, but that wasn't the key. The key was he felt connected.

He began calling relatives and other people he knew, talking, sharing, confiding. The connections seemed real as long as he was doped but disappeared as soon as the opioids wore off. Drug-manufactured intimacy, he learned, didn't last.

An intermittent pattern of opioid use followed Chris to his next attempt at matriculating at Stanford. When he returned in the fall of 2009, now his fourth attempt, he was chronologically and geographically marginalized from his undergraduate peers. He was five years older than the average sophomore.

He was placed in graduate student housing, where he shared a two-bedroom apartment with a graduate student in particle physics. They had little in common and worked hard to stay out of each other's way.

He developed a routine that revolved around studying and drug use. He had given up on the idea of trying to quit. He had come to think of himself as a confirmed "drug addict."

He smoked cannabis alone in his bedroom every day. Every Friday night he went up to San Francisco, alone, to get heroin. A single shot on the street cost him fifteen dollars, for a rush that lasted five to fifteen seconds, and an afterglow that persisted for hours. He smoked more cannabis to ease the comedown. Midway through the first quarter, he sold his laptop to buy more heroin. Then he sold his coat. He remembered being cold as he wandered the streets of the city.

He tried once to make friends with two British students in his language class. He told them he wanted to make a movie, with them in it. He had begun to take an interest in photography and sometimes wandered the campus taking pictures. They seemed initially charmed, but when he told them his idea for the movie—to film them speaking in American accents while eating—they demurred and avoided him thereafter.

"I guess I've always been odd like that. Odd ideas. That's why I don't ever want to tell people what I'm thinking."

Through it all, Chris went to class and got As, except one B in the Interpersonal Basis of Abnormal Behavior. He went home at Christmas and didn't return.

In the fall of 2010, Chris made one last half-hearted attempt to matriculate at Stanford. He rented a room off campus in Menlo Park and declared yet another new major: human biology. A few days in, he stole pain pills from his landlady and got a prescription for Ambien, which he crushed and injected. He made it five miserable months, then left Stanford with no hopes this time of ever returning.

Back home in Arkansas, Chris spent his days getting high. He would shoot up in the morning, and when it wore off hours later, he would lie in his bed in his parents' home, willing time to pass. The loop seemed endless and inescapable.

In the spring of 2011, Chris got caught by police stealing ice cream while intoxicated. He was offered jail or rehab. He chose rehab. On April 1, 2011, in rehab, Chris was started on a medication called buprenorphine, better known by the trade name Suboxone. Chris credits buprenorphine with saving his life.

After two years of stability on buprenorphine, Chris decided to make one final attempt at returning to Stanford. In 2013 he rented a bed in a trailer home from an elderly Chinese man. He couldn't afford anything else. In his first month on campus, he came to me looking for help.

Of course, I agreed to prescribe buprenorphine for Chris.

Three years later he graduated with honors and went on to get a PhD. His "odd" ideas, it turned out, were well suited to the laboratory.

In 2017, he married his girlfriend. She knew about his past and understood why he took buprenorphine. She sometimes lamented his "robotic lack of emotion," especially his apparent lack of anger when she felt anger was warranted.

But basically, life was good. Chris was no longer overwhelmed by craving, rage, and other intolerable emotions. He spent his days in the laboratory and rushed home after work to see his wife. They were soon expecting their first child.

One day in 2019, I said to Chris during one of our monthly sessions, "You're doing so well, and have been for so long, have you thought about trying to get off of buprenorphine?"

His answer was definitive. "I don't ever want to get off of buprenorphine. It was like a light switch for me. It didn't just prevent me from doing heroin. It gave my body something I needed and couldn't find anywhere else."

Medications to Restore a Level Balance?

I've thought often about what Chris said that day, about buprenorphine giving him something he couldn't find anywhere else.

Had prolonged drug use broken his pleasure-pain balance such that he would need opioids for the rest of his life just to feel "normal"? Perhaps some people's brains lose the plastic-ity necessary to restore homeostasis, even after prolonged abstinence. Perhaps even after the gremlins dismount, their balance remains permanently weighted to the side of pain.

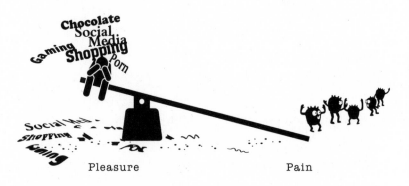

Pleasure   Pain

Or was Chris saying that opioids corrected a chemical im-balance he was born with?

When I went through medical school and residency in the 1990s, I was taught that people with depression, anxiety, at-tention deficit, cognitive distortions, sleep problems, and so on have brains that don't work the way they're supposed to, just like people with diabetes have a pancreas that doesn't se-crete enough insulin. My job, according to the theory, is to replace the missing chemical so people can function "nor-mally." This messaging was widely disseminated and aggres-sively promoted by the pharmaceutical industry and found a receptive audience in doctors and patient consumers alike.

Or maybe Chris was saying something different still. Maybe he was saying that buprenorphine made up for a deficit not in his brain, but in the world. Maybe the world let Chris down, and buprenorphine was the best way he could see to adapt.

Whether the problem was in Chris's brain or in the world, whether it was caused by prolonged drug use or a problem he was born with, here are some of the things I worry about in using medications to press on the pleasure side of the balance.

First, any drug that presses on the pleasure side has the potential to be addictive.

David, the college student who got hooked on prescription stimulants, is living proof that getting stimulants from a doctor for a diagnosed medical condition does not confer immunity to the problems of dependence and addiction. Prescription stimulants are the molecular equivalent of street methamphetamine (ice, speed, crank, Christina, no doze, Scooby snax). They cause a surge of dopamine in the brain's reward pathway and "have a high potential for abuse," a direct quote from the Food and Drug Administration's warning for Adderall.

Second, what if these drugs don't actually work the way they're supposed to, or worse yet, make psychiatric symptoms worse in the long run? Although buprenorphine was working for Chris, the evidence for psychotropic medications more generally is not robust, especially when taken long term.

Despite substantial increases in funding in four high-resourced countries (Australia, Canada, England, and the US) for psychiatric medications like antidepressants (Prozac), anxiolytics (Xanax), and hypnotics (Ambien), the prevalence of mood and anxiety symptoms in these countries has not decreased (1990 to 2015). These findings persist even when controlling for increases in risk factors for mental illness, such as poverty and trauma, and even when studying severe mental illness, such as schizophrenia.

Patients with anxiety and insomnia who take benzodiazepines (Xanax and Klonopin) and other sedative-hypnotics daily for more than a month may experience worsened anxiety and insomnia.

Patients with pain who take opioids daily for more than a month are at increased risk not only for opioid addiction but also for worsened pain. As mentioned earlier, this is the process called opioid-induced hyperalgesia, that is, opioids making pain worse with repeated doses.

Medications like Adderall and Ritalin prescribed for attention deficit disorder promote short-term memory and attention, but there is little or no evidence for enhanced long-term complex cognition, improved scholarship, or higher grades.

As public health psychologist Gretchen LeFever Watson and her co-authors wrote in *The ADHD Drug Abuse Crisis on American College Campuses*, "Compelling new evidence indicates that ADHD drug treatment is associated with deterioration in academic and social-emotional functioning."

Recent data show that even antidepressants, previously thought not to be "habit forming," may lead to tolerance and dependence, and possibly even make depression worse over the long haul, a phenomenon called *tardive dysphoria*.

Beyond the problem of addiction and the question of whether or not these drugs help, I've been plagued by a deeper question: What if taking psychotropic drugs is causing us to lose some essential aspect of our humanity?

In 1993, the psychiatrist Dr. Peter Kramer published his groundbreaking book *Listening to Prozac*, in which he argued that antidepressants make people "better than well." But what if Kramer got it wrong? What if instead of making us better than well, psychotropic drugs make us *other than well*?

I've had many patients over the years who have told me that their psychiatric medications, while offering short-term relief from painful emotions, also limit their ability to experience the full range of emotions, especially powerful emotions like grief and awe.

One patient who seemed to be doing well on antidepressants told me she no longer cried at Olympics commercials. She laughed when she talked about it, happily forfeiting the sentimental side of her personality for relief from depression and anxiety. But when she couldn't even cry at her own mother's funeral, the balance for her had tipped. She went off antidepressants and a short time later experienced a wider emotional amplitude, including more depression and anxiety. She decided the lows were worth it to feel human.

Another patient of mine who tapered off high-dose Oxy-Contin, which she'd taken for over a decade for chronic pain, came back to see me months later with her husband. It was my first time meeting him. He'd tired of so many doctors over so many years. "My wife on Oxy," he said, "stopped listening to music. Now off of that stuff she enjoys music again. For me it feels like I got back the person I married."

I've had my own experiences with psychotropic medication.

Restless and irritable from childhood, I was, for my mother, a difficult child to raise. She struggled to help me temper my moods and in the process felt bad about herself as a parent, or at least that's my interpretation of the past. She admits she preferred my brother, docile and biddable. I preferred him too, and he effectively raised me when my mother threw up her hands in frustration.

In my twenties, I started on Prozac for chronic low-grade irritability and anxiety diagnosed as "atypical depression." I felt better right away. Mostly, I stopped asking the big questions: *What is our purpose? Do we have free will? Why do we suffer? Is there a God?* Instead, I just sort of got on with it.

Also, for the first time in my life, my mother and I got along. She found me pleasant to be around, and I enjoyed being more pleasing. I fit her better.

When I went off Prozac some years later in anticipation of trying to get pregnant, I reverted to my old self: cranky, questioning, restless. Almost immediately, my mother and I were at odds again. The very air in the room seemed to crackle when we were both in it.

Our relationship decades later is marginally better. We do best when we interact least. This makes me sad because I love my mom and I know she loves me.

But I don't regret going off Prozac. My non-Prozac personality, although not a good fit for my mom, has allowed me to do things I never would have done otherwise.

Today, I'm finally okay with being a somewhat anxious, slightly depressed skeptic. I'm a person who needs friction, a challenge, something to work for or fight against. I won't whittle myself down to fit the world. Should any of us?

In medicating ourselves to adapt to the world, what kind of

world are we settling for? Under the guise of treating pain and mental illness, are we rendering large segments of the population biochemically indifferent to intolerable circumstance? Worse yet, have psychotropic medications become a means of social control, especially of the poor, unemployed, and disenfranchised?

Psychiatric drugs are prescribed more often and in larger amounts to poor people, especially poor children.

According to the 2011 data from the National Health Interview Survey of the CDC's National Center for Health Statistics, 7.5 percent of American children between the ages of six and seventeen took a prescribed medication for "emotional and behavioral difficulties." Poor children were more likely to take psychiatric medications than those not living in poverty (9.2 percent versus 6.6 percent). Boys were more likely than girls to be medicated. Non-Hispanic whites were more likely than people of color to be medicated.

Based on the extrapolation of Georgia Medicaid data to the rest of the nation, as many as ten thousand toddlers may be receiving psychostimulant medications like Ritalin.

As psychiatrist Ed Levin wrote regarding the problem of overdiagnosing and overmedicating American youth, especially among the poor: "While a tendency to rage must, as does all behavior, involve some biology, it may more significantly reflect a patient's reaction to adverse and inhumane treatment."

This phenomenon is not limited to the United States.

A nationwide study in Sweden analyzed rates of prescribing for different psychiatric drugs, based on indices of what they called "neighborhood deprivation" (index of education, income, unemployment, and welfare assistance). For each class of psychiatric medication, they found prescribing of psychiatric medications increased as the socioeconomic status of the neighborhood fell. Their conclusion: "These findings suggest that neighborhood deprivation is associated with psychiatric medication prescription."

Opioids too are disproportionately prescribed to the poor.

According to the US Department of Health and Human Services, "Poverty, unemployment rates, and the employment-to-population ratio are highly correlated with the prevalence of prescription opioids and with substance use measures. On average, counties with worse economic prospects are more likely to have higher rates of opioid prescriptions, opioid-related hospitalizations, and drug overdose deaths."

Americans on Medicaid, federally funded health insurance for the poorest and most vulnerable people, are prescribed opioid painkillers at twice the rate of non-Medicaid patients. Medicaid patients die from opioids at three to six times the rate of non-Medicaid patients.

Even medications like buprenorphine maintenance treatment (BMT), which is what I was prescribing to Chris to treat opioid addiction, may constitute a type of "clinical abandonment" when psychosocial determinants of health are not likewise addressed. As Alexandrea Hatcher and her colleagues wrote in the journal *Substance Use and Misuse*: "Without attention to the basic needs of patients without race and class privilege, BMT, as medication alone, rather than being liberatory, can turn into a form of institutional neglect and even structural violence to the extent that it is considered adequate for their recovery."

———————

The sci-fi movie *Serenity* (2005), directed by Joss Whedon, imagines a future world in which national leaders conduct a grand experiment: They inoculate an entire planet's population against greed, sadness, anxiety, anger, despair in hopes of achieving a civilization of peace and harmony.

Mal, a rogue pilot, the movie's hero, and the captain of the spaceship *Serenity*, travels with his crew to the planet to explore. Instead of finding Shangri-La, he finds corpses without a ready explanation for their death. An entire planet is dead in repose, lying in their beds, kicking back on their couches, slumped at their desks. Mal and his crew eventually puzzle it out: The genetic mutation deprived them of hunger for anything.

Like real-life dopamine-depleted rats who starve to death rather than shuffle a few centimeters for food, these humans died for lack of desire.

———————

Please don't misunderstand me. These medications can be lifesaving tools and I'm grateful to have them in clinical practice. But there is a cost to medicating away every type of human suffering, and as we shall see, there is an alternative path that might work better: embracing pain.

PART III

The Pursuit of Pain

CHAPTER 7

Pressing on the Pain Side

Michael sat across from me, looking relaxed in jeans and a T-shirt. Boyishly handsome and effortlessly charming, his natural appeal was both his gift and his burden.

"I'm an attention whore," he said. "Any of my friends will tell you that."

Michael's life was once upon a time a Silicon Valley fairy tale. After graduating from college, he made millions in the real estate business. By age thirty-five, he was fabulously rich, enviably handsome, and happily married to the woman he loved.

But he had another life that would soon unravel everything he had worked for.

"I've always been an energy guy, looking for anything to give me a boost. Cocaine was obvious, but alcohol did that for me too . . . gave me a euphoric high and lots of energy, from the very first time I tried it. I told myself I was going to be that one guy who could do cocaine recreationally and not get into trouble. At the time, I really believed that." He paused and smiled. "I should have known.

"When my wife told me that tackling my addiction was going to be the only way to save our marriage, I didn't even hesitate. I wanted her. I wanted the marriage. Recovery was the only choice."

Quitting, for Michael, wasn't the hard part. It was figuring out what to do next. After quitting he was flooded with all the negative emotions he'd been masking with drugs. When he wasn't feeling sad, angry, and ashamed, he was feeling nothing at all, which was possibly worse. Then he happened upon something that gave him hope.

"The first time it happened," he told me, "it was an accident. I'd been getting up in the mornings to take tennis lessons . . . a way to distract myself in the early days of not using. But an hour after tennis and showering, I'd still be sweating. I mentioned it to my tennis coach, and he suggested I try a cold shower instead. The cold shower was a little painful, but only for seconds until my body got used to it. When I got out, I felt surprisingly good, like I'd had a really good cup of coffee.

"Over the next couple of weeks, I started to notice that my mood after a cold shower was better. I researched cold-water therapy online and found a community of people taking ice baths. It seemed kind of crazy, but I was desperate. Following their lead, I progressed from cold showers to filling my bathtub with cold water and immersing myself in it. That worked even better, so I upped the ante and added ice to the tub water to get the temperature even lower. By doing that, I could get the temp to the mid-fifties.

"I got into a routine where I immersed myself in ice water for five to ten minutes every morning and again just before bed. I did that every day for the next three years. It was key to my recovery."

"What does it feel like," I asked, "immersing yourself in cold water?" I have an aversion to cold water myself, and couldn't tolerate those temperatures for even a few seconds.

"For the first five to ten seconds, my body is screaming: *Stop, you're killing yourself.* It's that painful."

"I can imagine."

"But I tell myself it's time limited, and it's worth it. After the initial shock, my skin goes numb. Right after I get out, I feel high. It's exactly like a drug . . . like how I remember ecstasy or recreational Vicodin. Incredible. I feel great for hours."

For most of human history, people bathed in cold water. Only those living near a natural hot spring could regularly enjoy a hot bath. No wonder people back then stayed dirtier.

The ancient Greeks developed a heating system for public baths but continued to advocate for the use of cold water to treat a variety of ailments. In the 1820s, a German farmer named Vincenz Priessnitz promoted the use of ice-cold water to cure all manner of physical and psychological disorders.

He went so far as to turn his home into a sanitarium for ice-water treatment.

Since the advent of modern plumbing and heating, hot baths and showers have become the norm; but ice-water immersion has lately become popular again.

Endurance athletes claim it speeds muscle recovery. The "Scottish shower," also called the "James Bond shower" as practiced by James Bond in Ian Fleming's 007 novels, is newly popular and consists of ending a hot shower with at least a minute of cold shower.

Ice-water immersion gurus such as the Dutchman Wim Hof have become celebrities in their own right for their ability to immerse themselves for hours at a time in near-freezing temperatures.

Scientists at Charles University in Prague, writing in the *European Journal of Applied Physiology*, conducted an experiment in which ten men volunteered to submerge themselves (head out) in cold water (14 degrees Celsius) for one hour. This is 57 degrees Fahrenheit.

Using blood samples, the researchers showed that plasma (blood) dopamine concentrations increased 250 percent, and plasma norepinephrine concentrations increased 530 percent as a result of cold-water immersion.

Dopamine rose gradually and steadily over the course of the cold bath and remained elevated for an hour afterward. Norepinephrine rose precipitously in the first thirty minutes, plateaued in the latter thirty minutes, and dropped by about a third in the hour afterward, but it remained elevated well above baseline even into the second hour after the bath. Dopamine and norepinephrine levels endured well beyond the painful stimulus itself, which explains Michael's statement, "Right after I get out . . . I feel great for hours."

Other studies examining the brain effects of cold-water immersion in humans and animals show similar elevations in monoamine neurotransmitters (dopamine, norepinephrine, serotonin), the same neurotransmitters that regulate pleasure, motivation, mood, appetite, sleep, and alertness.

Beyond neurotransmitters, extreme cold in animals has been shown to promote neuronal growth, all the more remarkable since neurons are known to alter their microstructure in response to only a small handful of circumstances.

Christina G. von der Ohe and her colleagues studied the brains of hibernating ground squirrels. During hibernation, both core and brain temperatures drop to within 0.5–3 degrees Celsius. At freezing temperatures, the neurons of hibernating ground squirrels look like spindly trees with few branches (dendrites) and even fewer leaves (microdendrites).

As the hibernating ground squirrel is warmed, however, the neurons show remarkable regrowth, like a deciduous forest at the height of spring. This regrowth occurs rapidly, rivaling the kind of neuronal plasticity seen only in embryonic development.

The study's authors wrote of their findings: "The structural changes we have demonstrated in the hibernator brain are among the most dramatic found in nature. . . . Whereas dendritic elongation can reach 114 micrometers per day in the hippocampus of the developing rhesus monkey embryo, adult hibernators exhibit similar changes in just 2 hours."

Michael's accidental discovery of the benefits of ice-cold water immersion is an example of how pressing on the pain side of the balance can lead to its opposite—pleasure. Unlike pressing on the pleasure side, the dopamine that comes from pain is indirect and potentially more enduring. So how does it work?

Pain leads to pleasure by triggering the body's own regulating homeostatic mechanisms. In this case, the initial pain stimulus is followed by gremlins hopping on the pleasure side of the balance.

The pleasure we feel is our body's natural and reflexive

Pleasure   Pain

physiological response to pain. Martin Luther's mortification of the flesh through fasting and self-flagellation may have gotten him a little bit high, even if it was for religious reasons.

With intermittent exposure to pain, our natural hedonic set point gets weighted to the side of pleasure, such that we become less vulnerable to pain and more able to feel pleasure over time.

Pleasure   Pain

In the late 1960s, scientists conducted a series of experiments on dogs that, due to the experiments' obvious cruelty, would not be allowed today but nonetheless provide important information on brain homeostasis (or leveling the balance).

After connecting the dog's hind paws to an electrical current, the researchers observed: "The dog appeared to be terrified during the first few shocks. It screeched and thrashed about, its pupils dilated, its eyes bulged, its hair stood on end, its ears lay back, its tail curled between its legs.

Expulsive defecation and urination, along with many other symptoms of intense autonomic nervous system activity, were seen."

After the first shock, when the dog was freed from the harness, "it moved slowly about the room, appeared to be stealthy, hesitant, and unfriendly." The dog's heart rate increased to 150 beats per minute above resting baseline during the first shock. When the shock was over, the dog's heart rate slowed to 30 beats below baseline for a full minute.

Over subsequent electric shocks, "its behavior gradually changed. During shocks, the signs of terror disappeared. Instead, the dog appeared pained, annoyed, or anxious, but not terrified. For example, it whined rather than shrieked, and showed no further urination, defecation, or struggling. Then, when released suddenly at the end of the session, the dog rushed about, jumped up on people, wagged its tail, in what we called at the time 'a fit of joy.'"

With subsequent shocks, the dog's heart rate rose only slightly above resting baseline, and then only for a few seconds. After the shock was over, the heart rate slowed massively to 60 beats per minute below resting baseline, double the first time. It took a full five minutes for the heart rate to return to resting baseline.

With repeated exposure to a painful stimulus, the dog's mood and heart rate adapted in kind. The initial response (pain) got shorter and weaker. The after-response (pleasure) got longer and stronger. Pain morphed into hypervigilance morphed into a "fit of joy." An elevated heart rate, consistent with a fight-or-flight reaction, morphed into minimal heart rate elevation followed by prolonged bradycardia, a slowed heart rate seen in states of deep relaxation.

It's not possible to read this experiment without feeling pity for the animals subjected to this torture. Yet the so-called "fit of joy" suggests a tantalizing possibility: By pressing on the pain side of the balance, might we achieve a more enduring source of pleasure?

This idea is not new. Ancient philosophers observed a similar phenomenon. Socrates (as recorded by Plato in "Socrates' Reasons for Not Fearing Death") mused on the relationship between pain and pleasure more than two thousand years ago:

> How strange would appear to be this thing that men call pleasure! And how curiously it is related to what is thought to be its opposite, pain! The two will never be found together in a man, and yet if you seek the one and obtain it, you are almost bound always to get the other as well, just as though they were both attached to one and the same head. . . . Wherever the one is found, the other follows up behind. So, in my case, since I had pain in my leg as a result of the fetters, pleasure seems to have come to follow it up.

The American cardiologist Helen Taussig published an article in *American Scientist* in 1969 in which she described the experiences of people struck by lightning who lived to tell about it. "My neighbor's son was struck by lightning as he was returning from a golf course. He was thrown to the ground. His shorts were torn to shreds and he was burned across his thighs. When his companion sat him up, he screamed 'I'm dead, I'm dead.' His legs were numb and blue and he could not move. By the time he reached the nearest hospital he was euphoric. His pulse was very slow." This account recalls the dog's "fit of joy," including the slowed pulse.

We've all experienced some version of pain giving way to pleasure. Perhaps like Socrates, you've noticed an improved mood after a period of being ill, or felt a runner's high after exercise, or took inexplicable pleasure in a scary movie. Just as pain is the price we pay for pleasure, so too is pleasure our reward for pain.

The Science of Hormesis

Hormesis is a branch of science that studies the beneficial effects of administering small to moderate doses of noxious and/or painful stimuli, such as cold, heat, gravitational changes, radiation, food restriction, and exercise. *Hormesis* comes from the ancient Greek *hormáein*: to set in motion, impel, urge on.

Edward J. Calabrese, an American toxicologist and a leader in the field of hormesis, describes this phenomenon as the "adaptive responses of biological systems to moderate environmental or self-imposed challenges through which the system improves its functionality and/or tolerance to more severe challenges."

Worms exposed to temperatures above their preferred 20 degrees Celsius (35 degrees C for two hours) lived 25 percent longer and were 25 percent more likely to survive subsequent high temperatures than nonexposed worms. But too much heat wasn't good. Four hours as opposed to two hours of heat exposure reduced subsequent heat tolerance and reduced lifespan by a fourth.

Fruit flies that were spun in a centrifuge for two to four weeks not only outlived unspun flies but were also more agile in their older age, able to climb higher and longer than their nonexposed counterparts. But flies spun longer than that did not thrive.

Among Japanese citizens living outside the epicenter of the 1945 nuclear attack, those with low-dose radiation exposure may have shown marginally longer lifespans and decreased rates of cancer compared to un-irradiated individuals. Of those living in the direct vicinity of the atomic blast, approximately 200,000 died instantaneously.

The authors theorized that "low-dose stimulation of DNA damage repair, the removal of aberrant cells via stimulated apoptosis [cell death], and elimination of cancer cells via stimulated anticancer immunity" are at the heart of the beneficial effects of radiation hormesis.

Note that these findings are controversial, and a follow-up paper published in the prestigious *Lancet* disputed them.

Intermittent fasting and calorie restriction extended lifespan and increased resistance to age-related diseases in rodents and monkeys, as well as reduced blood pressure and increased heart rate variability.

Intermittent fasting has become somewhat popular as a way to lose weight and improve well-being. Fasting algorithms include alternate-day fasting, one-day-per-week fasting, up-to-the-ninth-hour fasting, one-meal-per-day fasting, 16:8 fasting

(fasting for sixteen hours each day and doing all your eating within the other eight-hour window), and so on.

American celebrity talk show host Jimmy Kimmel practices intermittent fasting. "Something I've been doing for a couple of years now is starving myself two days a week. . . . On Monday and Thursday, I eat fewer than five hundred calories a day, then I eat like a pig for the other five days. You 'surprise' the body, keep it guessing."

Not long ago, such fasting behaviors might have warranted the label "eating disorder." Too few calories is harmful for obvious reasons. But today, fasting in some circles is considered normal and even healthy.

What about exercise?

Exercise is immediately toxic to cells, leading to increased temperatures, noxious oxidants, and oxygen and glucose deprivation. Yet the evidence is overwhelming that exercise is health-promoting, and the absence of exercise, especially combined with chronic sedentary feeding—eating too much all day long—is deadly.

Exercise increases many of the neurotransmitters involved in positive mood regulation: dopamine, serotonin, norepinephrine, epinephrine, endocannabinoids, and endogenous opioid peptides (endorphins). Exercise contributes to the birth of new neurons and supporting glial cells. Exercise even reduces the likelihood of using and getting addicted to drugs.

When rats were given access to a running wheel six weeks

prior to gaining free access to cocaine, they self-administered the cocaine later and less often than rats who had not had prior wheel training. This finding has been replicated with heroin, methamphetamine, and alcohol. When exercise is not voluntary but rather forced on the animal, it still results in reduced voluntary drug consumption.

In humans, high levels of physical activity in junior high, high school, and early adulthood predict lower levels of drug use. Exercise has also been shown to help those already addicted to stop or cut back.

Dopamine's importance to motor circuits has been reported for every animal phylum in which it has been investigated. The nematode *C. elegans*, a worm and one of the simplest laboratory animals, releases dopamine in response to environmental stimuli signaling the local abundance of food. Dopamine's ancient role in physical movement relates to its role in motivation: To obtain the object of our desire, we need to go get it.

Of course today's easy-access dopamine doesn't require us to get off the couch. According to survey reports, the typical American today spends half their waking hours sitting, 50 percent more than fifty years ago. Data from other rich nations around the globe are comparable. When you consider that we evolved to traverse tens of kilometers daily to compete for a limited supply of food, the adverse effects of our modern sedentary lifestyle are devastating.

I sometimes wonder if our modern predilection for becoming addicted is fueled in part by the way drugs remind us that we still have bodies. The most popular video games feature avatars that run, jump, climb, shoot, and fly. The smartphone requires us to scroll through pages and tap on screens, cleverly exploiting ancient habits of repetitive motion, possibly acquired through centuries of grinding wheat and picking berries. Our contemporary preoccupation with sex may be because it's the last physical activity still widely practiced.

A key to well-being is for us to get off the couch and move our real bodies, not our virtual ones. As I tell my patients, just walking in your neighborhood for thirty minutes a day can make a difference. That's because the evidence is indisputable: Exercise has a more profound and sustained positive effect on mood, anxiety, cognition, energy, and sleep than any pill I can prescribe.

But pursuing pain is harder than pursuing pleasure. It goes against our innate reflex to avoid pain and pursue pleasure. It adds to our cognitive load: We have to *remember* that we will feel pleasure after pain, and we're remarkably amnestic about this sort of thing. I know I have to relearn the lessons of pain every morning as I force myself to get out of bed and go exercise.

Pursuing pain instead of pleasure is also countercultural, going against all the feel-good messages that pervade so many aspects of modern life. Buddha taught finding the Middle Way between pain and pleasure, but even the Middle Way has been adulterated by the "tyranny of convenience."

Hence we must seek out pain and invite it into our lives.

Pain to Treat Pain

The intentional application of pain to treat pain has been around since at least Hippocrates, who wrote in his *Aphorisms* in 400 BC: "Of two pains occurring together, not in the same part of the body, the stronger weakens the other."

The history of medicine is replete with examples of using painful or noxious stimuli to treat painful disease states. Sometimes called "heroic therapies"—cupping, blisters, cauterizing, moxibustion—painful remedies were widely practiced prior to 1900. The popularity of heroic therapies began to decline in the twentieth century as the medical profession discovered drug therapy.

With the advent of pharmacotherapy, pain to treat pain came to be seen as a kind of quackery. But as the limitations and harms of pharmacotherapy have moved to the forefront in recent decades, there has been a resurgence of interest in nonpharmacologic therapies, including painful remedies.

In 2011, in an article in a leading medical journal, Christian Sprenger and his colleagues from Germany provided empirical support for Hippocrates's ancient ideas about pain. They used neuroimaging (pictures of the brain in real time) to study the effects of heat and other painful stimuli applied to the arms and legs of twenty healthy young men.

They found that the subjective experience of pain caused by an initial painful stimulus was lessened with the application of a second painful stimulus. Further, naloxone, an opioid receptor blocker, prevented this phenomenon, suggesting that the application of pain triggers the body's own endogenous (self-made) opioids.

Liu Xiang, a professor at the China Academy of Traditional Chinese Medicine in Beijing, published a paper in 2001 in the *Chinese Science Bulletin,* revisiting the centuries-old practice of acupuncture and relying on modern science to explain how it works. He argued that the efficacy of acupuncture is mediated through pain, with needle insertion as the primary mechanism: "The needling, which can injure the tissue, is a noxious stimulation inducing pain . . . inhibiting great pain with little pain!"

The opioid receptor blocker naltrexone is currently being explored as a medical treatment for chronic pain. The idea is that by blocking the effects of opioids, including the ones we make (endorphins), we trick our bodies into making more opioids as an adaptive response.

Twenty-eight women with fibromyalgia took one pill of low-dose naltrexone (4.5 milligrams) a day for twelve weeks, and a sugar pill (placebo) for four weeks. Fibromyalgia is a chronic pain condition of unknown etiology thought possibly to be related to an individual's innate lower threshold for tolerating pain.

The study was double-blind, meaning that neither the women participating in the study nor the health care team knew which pill they were taking. Each woman was given a handheld computer to record her pain, fatigue, and other symptoms on a daily basis, and they continued to record their symptoms for four weeks after they stopped taking the capsules.

The study's authors reported that "[p]articipants experienced a significantly greater reduction in their pain scores while they were taking the LDN [low-dose naltrexone] as compared with placebo. They also reported improved general satisfaction with life and improved mood while taking LDN."

———

Electricity applied to the brain to treat mental illness has been practiced since the early 1900s. In April 1938, Ugo Cerletti and Lucino Bini performed the first electroconvulsive shock therapy (ECT) treatment on a forty-year-old patient whom they described as follows:

"He expressed himself exclusively in an incomprehensible gibberish made up of odd neologisms and, since his arrival from Milan by train without a ticket, not a thing had been ascertainable about his identity."

When Cerletti and Bini applied electricity to his brain for the first time, they observed "a sudden jump of the patient on his bed with a very short tensing of all his muscles; then he immediately collapsed onto the bed without loss of consciousness. The patient presently started to sing at the top of his voice, then fell silent. It was evident from our experience with dogs that the voltage had been held too low."

Cerletti and Bini argued as to whether they should apply yet another shock at a higher voltage. While they were talking, the patient cried out, "Non una seconda! Mortifera!" ("Not again! It will kill me!"). Despite his protests, they applied a second shock—a cautionary tale against arriving in Milan without a train ticket or "ascertainable identity" in 1938.

Once the "patient" had recovered from the second shock, Cerletti and Bini observed he "sat up of his own accord, looked about him calmly with a vague smile, as though asking what was expected of him. I asked him 'what has been happening to you?' He answered, with no more gibberish: 'I don't know, perhaps I have been asleep.' The initial patient received thirteen more ECT treatments over two months and was, per report, discharged in complete recovery."

ECT is still practiced today to good effect, although much more humanely. Muscle relaxants and paralytics prevent painful contractions. Anesthetics allow patients to remain asleep and mostly unconscious throughout the procedure. So it cannot be said today that pain per se is the mediating factor.

Nonetheless, ECT provides a hormetic shock to the brain, which in turn spurs a broad compensatory response to reassert homeostasis: "ECT brings about various neuro-physiological as well as neuro-chemical changes in the macro- and micro-environment of the brain. Diverse changes involving expression of genes, functional connectivity, neurochemicals, permeability of blood-brain-barrier, alteration in immune system has [sic] been suggested to be responsible for the therapeutic effects of ECT."

You'll remember David, the shy computer buff who ended up in the hospital after getting addicted to prescription stimulants.

After he was discharged, he began weekly exposure therapy

with a talented young therapist on our team. The basic principle of exposure therapy is to expose people in escalating increments to the very thing—being in crowds, driving across bridges, flying in airplanes—that causes the uncomfortable emotion they're trying to flee, and in doing so, augment their ability to tolerate that activity. In time they may even come to enjoy it.

As the philosopher Friedrich Nietzsche famously said, a sentiment echoed by many before and after through the ages, "What doesn't kill me makes me stronger."

Given that David's greatest fear was talking to strangers, his first task was to force himself to make small talk with coworkers.

"My therapy homework," he told me months later, "was to go to the kitchen, the break room, or the cafeteria at work and talk to random people. I had a script: 'Hi. My name is David. I work in software development. What do you do?' I set a schedule: before lunch, at lunch, and after lunch. Then I had to measure my distress before, during, and after, on a scale from one to one hundred, with one hundred being the worst distress I could imagine."

In a world where we're increasingly counting ourselves— steps, breaths, heartbeats—putting a number on something has become a way we both master and describe experience. For me, quantifying things is not second nature, but I've learned to adapt, since this method of self-awareness seems to resonate especially well for the science-minded computer and engineering types we have so many of here in Silicon Valley.

"How did you feel before the interaction? Uh, what number were you?" I asked.

"Before I was one hundred. I just felt so terrified. My face got all red. I was sweating."

"What were you afraid would happen?"

"I was afraid of other people looking at me and laughing. Or calling human relations or security on me, because I seemed crazy."

"How did it go?"

"None of the things I was afraid would happen, happened. No one called HR or security. I stayed in the moment as long as possible, just letting my anxiety wash over me, while also being respectful of their time. The interactions lasted maybe four minutes."

"How did you feel afterward?"

"I was about a forty afterward. Much less anxious. So I did that on a schedule three times a day for weeks, and progressively over time it got easier and easier. Then I challenged myself with people outside of work."

"Tell me."

"At Starbucks, I intentionally made small talk with the barista. I never would have done that in the past. I always ordered with the app to avoid having to interact with a person at all. But this time, I went right up to the counter and ordered my coffee. My biggest fear was saying or doing something stupid. I was doing fine until I spilled a little bit of my coffee on the counter. I was so embarrassed. When I told my therapist about it, she told me to do it again—spill my coffee—on purpose this time. The next time I was at Starbucks, I spilled my coffee on purpose. I felt anxious, but I got used to it."

"What are you smiling about?"

"I almost can't believe how different my life is now. I'm less on guard. I don't have to preplan so much to avoid interacting with people. I can get on a crowded train now and not be late for work because I wait for the next one, and the one after that. I actually enjoy meeting people I'll never see again."

Alex Honnold, now world-famous for climbing the face of Yosemite's El Capitan without ropes, was found to have below-normal amygdala activation during brain imaging. For most of us, the amygdala is an area of the brain that lights up in an fMRI machine when we look at scary pictures.

The researchers who studied Honnold's brain speculated that he was born with less innate fear than others, which in turn allowed him, they hypothesized, to accomplish superhuman climbing feats.

But Honnold himself disagreed with their interpretation: "I've done so much soloing, and worked on my climbing skills so much that my comfort zone is quite large. So these things that I'm doing that look pretty outrageous, to me they seem normal."

The most likely explanation for Honnold's brain differences is the development of tolerance to fear through neuroadaptation. My guess is that Honnold's brain started out no different from the average brain in terms of fear sensitivity. What's different now is that he has trained his brain through years of climbing not to react to fearful stimuli. It takes a lot more to scare Honnold's brain than the average person's because he has incrementally exposed himself to death-defying feats.

Of note, Honnold nearly had a panic attack when he went inside the fMRI machine to get pictures taken of his "fearless brain," which also tells us that fear tolerance doesn't necessarily translate across all experiences.

Alex Honnold and my patient David have been climbing different parts of the same fear mountain. Just as Honnold's brain adapted to climbing a rock face without ropes, David developed the mental calluses that made him able to tolerate anxiety, and gained a sense of confidence and competence about himself and his capacity to live in the world.

Pain to treat pain. Anxiety to treat anxiety. This approach is counterintuitive, and exactly opposite to what we've been taught over the last 150 years about how to manage disease, distress, and discomfort.

Addicted to Pain

"Over time I realized the more pain I felt with the initial shock of cold water," said Michael, "the bigger the high afterward. So I started to find ways to up the ante.

"I bought a meat freezer—a trough with a lid and built-in cooling coils—and filled it with water every night. By morning, there was a thin layer of ice on the surface, with temperatures in the low thirties. Before getting in, I had to break ice.

"Then I read that the body heats up the water after a few

minutes, unless the water is moving, like a whirlpool. So I bought a motor to go into the ice bath. That way, I could sustain near-freezing temperatures while I was in it. I also bought a hydropowered mattress pad for my bed, which I keep at the lowest temperatures, about 55°F (13°C)."

Michael stopped talking abruptly and looked at me with a lopsided smile. "Wow. I realize as I'm talking about this . . . it sounds like an addiction."

In April 2019, Professor Alan Rosenwasser of the University of Maine e-mailed me, looking for a copy of a chapter I had recently published with a colleague on the role of exercise in treating addiction. He and I had never met. After getting permission from the publisher, I sent him the chapter.

Approximately a week later he wrote again, this time with the following.

> Thanks for sharing. One issue I notice that you did not discuss is the question of whether wheel-running in mice and rats is a model for voluntary exercise or for pathological exercise (exercise addiction). Some animals housed in wheels exhibit what might be considered excessive levels of running, and one study has shown that wild rodents will use a running wheel that has been left outside in the environment.

I was fascinated and wrote him back immediately. What followed was a series of conversations in which Dr. Rosenwasser, who has spent the last forty years studying circadian rhythms, also known as the "clocks field," schooled me in running wheels.

"When people first started doing this work," Rosenwasser told me, "it was assumed, mistakenly, that running wheels were a way to keep track of the animals' spontaneous activity: rest versus movement. Somewhere along the way, people became sensitive to the fact that running wheels are not inert. They're interesting in themselves. One of the kickstarters was adult hippocampal neurogenesis."

This refers to the discovery some decades ago that contrary to previous teaching, humans can generate new neurons in the brain into middle and late adulthood.

"Once people accepted that new neurons are born and integrated into neural circuitry," Rosenwasser continued, "one of the easiest ways to stimulate neurogenesis was with a running wheel, even more potent than enriched environments [complex mazes, for example]. This led to a whole era of running wheel research.

"It turns out," Rosenwasser said, "that running wheels are governed by the same endo-opioid, dopamine, endo-cannabinoid pathways that drive compulsive drug use. It's important to know that running wheels are not necessarily a model for a healthy lifestyle."

In short, running wheels are a drug.

Mice placed in a complex maze of 230 meters of tunnels, including water, food, digging material, nests—in other words, a big area with a lot of cool stuff to do—as well as a running wheel, will spend much of their time on the running wheel and leave large segments of the maze unexplored.

Once rodents start using a running wheel, it's hard for them to stop. Rodents run much farther on a running wheel than they do on a flat treadmill or in a maze, and also much farther than they do during normal locomotion in natural environments.

Caged rodents given access to a running wheel will run until their tails are permanently curved upward and back toward their heads in the shape of the running wheel: the smaller the wheel, the sharper the curve of the tail. In some cases, rats run until they die.

The location, novelty, and complexity of the running wheel influence its use.

Wild mice prefer square wheels to circular ones, and wheels with hurdles contained within them to wheels without hurdles. They display a remarkable amount of coordination and acrobatic skill in running wheels. Like teens in a skateboard park, they allow "themselves to be repeatedly carried nearly to the top of the wheel in both forward and backward directions, running on the outside of the wheel on the top surface, or 'up' the outside of the wheel while balanced on their tail."

C. M. Sherwin in his 1997 review of running wheels speculated on the intrinsic reinforcing properties of running wheels:

> The three-dimensional quality of wheel running may be reinforcing to animals. During wheel running, an animal

will experience rapid changes in the speed and direction of its motion, owing in part to exogenous forces: the momentum and inertia of the wheel. This experience may be reinforcing, analogous with (some!) humans enjoying pleasure rides at the fairground, particularly for motion in the vertical plane . . . such changes in the motion of the animal are unlikely to be experienced in "natural" circumstances.

Johanna Meijer and Yuri Robbers of Leiden University Medical Center in the Netherlands put a running wheel in an urban area where feral mice live, and another in a dune not accessible to the public. They placed a video camera in each site to record every animal who visited the cages over two years.

The result was hundreds of instances of animals using the running wheels. "The observations showed that feral mice ran in the wheels year-round, steadily increasing in late spring and peaking in summer in the green urban area, while increasing in mid-to-late summer in the dunes, reaching a peak late in autumn."

Use of the wheel was not limited to wild mice. There were also shrews, rats, snails, slugs, and frogs, most of whom demonstrated intentional and purposeful engagement with the wheel.

The authors concluded that "wheel running can be experienced as rewarding even without an associated food reward, suggesting the importance of motivational systems unrelated to foraging."

Extreme sports—skydiving, kitesurfing, hang gliding, bob-sledding, downhill skiing/snowboarding, waterfall kayaking, ice climbing, mountain biking, canyon swinging, bungee jumping, base jumping, wingsuit flying—slam down hard and fast on the pain side of the pleasure-pain balance. Intense pain/fear plus a shot of adrenaline creates a potent drug.

Scientists have shown that stress alone can increase the release of dopamine in the brain's reward pathway, leading to the same brain changes seen with addictive drugs like cocaine and methamphetamine.

Just as we become tolerant to pleasure stimuli with repeated exposure, so too can we become tolerant to painful stimuli, resetting our brains to the side of pain.

A study of skydivers compared to a control group (rowers) found that repeat skydivers were more likely to experience anhedonia, a lack of joy, in the rest of their lives.

The authors wrote that "skydiving has similarities with addictive behaviors and that frequent exposure to 'natural high' experiences is related to anhedonia." I would hardly call jumping out of an airplane at 13,000 feet a "natural high," but I do agree with the author's overall conclusion: Skydiving can be addictive and can lead to persistent dysphoria if engaged in repeatedly.

Technology has allowed us to push the limits of human pain.

On July 12, 2015, ultramarathoner Scott Jurek broke the

speed record for running the Appalachian Trail. He ran from Georgia to Maine (2,189 miles) in 46 days, 8 hours, and 7 minutes. To accomplish this feat, he relied on the following technology and devices: lightweight, waterproof, heatproof clothing, "air-mesh" running shoes, a GPS satellite tracker, a GPS watch, an iPhone, hydration systems, electrolyte tablets, aluminum foldable trekking poles, "industrial water sprayers to simulate misting," "an ice cooler to cool my core down," 6,000–7,000 calories a day, and a pneumatic compression leg-massaging machine powered by solar panels on top of his support van, driven by his wife and crew.

In November 2017, Lewis Pugh swam a kilometer in $-3°$C ($26°$F) water near Antarctica in nothing but his swimsuit. Getting there required travel by air and sea from Pugh's native South Africa to South Georgia, a remote British island. As soon as Pugh was done swimming, his crew whisked him to a nearby ship, where he was immersed in hot water and where he remained for the next fifty minutes, to bring his core body temperature back to normal. Without this intervention, he surely would have died.

Alex Honnold's ascent of El Capitan seems like the ultimate technology-eschewing human accomplishment. No ropes. No gear. Just one person against gravity in a death-defying display of courage and mastery. But by all accounts, Honnold's feat would not have been possible without the "hundreds of hours on Freerider [the route he took], attached to ropes, working out a precisely rehearsed choreography for each section, memorizing thousands of intricate hand and foot sequences."

Honnold's ascent was filmed by a professional film crew and turned into a movie watched by millions, leading to a massive social media following and worldwide fame. Riches and celebrity, another dimension of our dopamine economy, contribute to the addictive potential of these extreme sports.

"Overtraining syndrome" is a well-described but poorly understood condition among endurance athletes who train so much that they reach a point where exercise no longer produces the endorphins that were once so plentiful. Instead, exercise leaves them feeling depleted and dysphoric, as if their reward balance has maxed out and stopped working, similar to what we saw with my patient Chris and opioids.

I'm not suggesting that everyone who engages in extreme and/or endurance sports is addicted, but rather highlighting that the risk of addiction to any substance or behavior increases with increasing potency, quantity, and duration. People who lean too hard and too long on the pain side of the balance can also end up in a persistent dopamine deficit state.

Too much pain, or in too potent a form, can increase the risk of becoming addicted to pain, something I've witnessed in clinical practice. A patient of mine ran so much she developed fractures in her leg bones and even then didn't stop running. Another patient cut her inner forearms and thighs with a razor blade to feel a rush and to calm the constant ruminations of her mind. She couldn't stop cutting even at the risk of severe scarring and infection.

When I conceptualized their behaviors as addictions and treated them as I would any patient with addiction, they got better.

Addicted to Work

The "workaholic" is a celebrated member of society. Nowhere is that perhaps more true than here in Silicon Valley, where 100-hour workweeks and 24/7 availability are the norm.

In 2019, after three years of traveling monthly for work, I decided to limit travel in an effort to bring work and home life back into balance. At first I transparently let people know the reason why: I wanted more time with my family. People seemed both annoyed and offended that I would decline their invitation for a reason as hippie-dippie as "time with family." I eventually resorted to saying I had another engagement, which was met with less resistance. My working elsewhere, it seemed, was acceptable.

Invisible incentives are now woven into the fabric of white-collar work, from the prospect of bonuses and stock options to the promise of promotion. Even in fields like medicine, health care providers see more patients, write more prescriptions, and perform more procedures, because they're incentivized to do so. I get a monthly report on my productivity, as measured by how much I've billed on behalf of my institution.

By contrast, blue-collar jobs are increasingly mechanized and divorced from the meaning of the work itself. Working under the employ of distant beneficiaries, there's limited autonomy, modest financial gain, and little sense of common mission. Piecemeal assembly-line work fragments the sense of accomplishment and minimizes contact with the end-product consumer, both of which are central to internal motivation. The result is a "work-hard/play-hard" mentality in which compulsive overconsumption becomes the reward at the end of a day of drudgery.

It's no wonder, then, that those with less than a high school education in low-paying jobs are working less than ever, whereas highly educated wage earners are working more.

By 2002, the top-paid 20 percent were twice as likely to work long hours as the lowest-paid 20 percent, and that trend continues. Economists speculate that this change is due to higher rewards for those at the top of the economic food chain.

I find it difficult at times to stop myself from working once I've begun. The "flow" of deep concentration is a drug in itself, releasing dopamine and creating its own high. This kind of single-minded focus, although heavily rewarded in modern rich nations, can be a trap when it keeps us from the intimate connections with friends and family in the rest of our lives.

The Verdict on Pain

As if answering his own question about whether he had gotten addicted to cold-water immersion, Michael said, "It never got out of hand. For two to three years, I took a ten-minute ice bath every morning. Now I'm not as into it as I used to be. I do it on average three times a week.

"What's really cool," he went on, "is that it's become a family activity, and something we do with friends. Doing drugs was always social. In college a lot of people partied hard. It was always sitting around together drinking or doing lines of coke.

"Now I don't do that anymore. Instead, a couple of our friends come over. . . . They have kids too, and we have a cold-water party. I have a custom trough set in the mid-forties, and everybody takes turns getting in, alternating with the hot tub. We have a timer and we cheer each other on, including the kids. The trend has caught on among our friends too. This group of all women in our friend group goes to the Bay once a week and gets in. They immerse themselves to their necks. That water's in the fifties."

"Then what?"

"I don't know," he laughed. "They probably go out and party." We both smiled.

"You've said several times you do it because it makes you feel alive. Can you explain?"

"I don't really like the feeling of being alive. Drugs and alcohol were a way to like it. Now I can't do that anymore. When I see people partying, I'm still a little bit jealous of the escape they're getting. I can see they get the reprieve. Cold water reminds me that being alive can feel good."

———

If we consume too much pain, or in too potent a form, we run the risk of compulsive, destructive overconsumption.

But if we consume just the right amount, "inhibiting great pain with little pain," we discover the path to hormetic healing, and maybe even the occasional "fit of joy."

CHAPTER 8

Radical Honesty

Every major religion and code of ethics has included honesty as essential to its moral teachings. All my patients who have achieved long-term recovery have relied on truth-telling as critical for sustained mental and physical health. I too have become convinced that radical honesty is not just helpful for limiting compulsive overconsumption but also at the core of a life well lived.

The question is, how does telling the truth improve our lives?

Let's first establish that telling the truth is painful. We're wired from the earliest ages to lie, and we all do it, whether or not we care to admit it.

Children begin lying as early as age two. The smarter the kid, the more likely they are to lie, and the better they are at it. Lying tends to decrease between ages three and fourteen, possibly because children become more aware of how lying harms other people. On the other hand, adults are capable of more sophisticated antisocial lies than children, as the ability to plan and remember becomes more advanced.

The average adult tells between 0.59 and 1.56 lies daily. *Liar, liar, pants on fire.* We've all got a little smoke coming off our shorts.

Humans are not the only animals with the capacity for deception. The animal kingdom is rife with examples of deception as a weapon and a shield. The *Lomechusa pubicollis* beetle, for example, is able to penetrate ant colonies by pretending to be one of them, something it accomplishes by emitting a chemical substance that makes it smell like an ant. Once inside, the beetle feeds on ant eggs and larvae.

But no other animal rivals the human capacity for lying.

Evolutionary biologists speculate that the development of human language explains our tendency and superior ability to lie. The story goes like this. The evolution of *Homo sapiens* culminated in the formation of large social groups. Large social groups were possible because of the development of sophisticated forms of communication, allowing for advanced mutual cooperation. Words used to cooperate can also be used to deceive and misdirect. The more advanced the language, the more sophisticated the lies.

Lies arguably have some adaptive advantage when it comes to competing for scarce resources. But lying in a world of plenty risks isolation, craving, and pathological overconsumption. Let me explain.

———————

"You look well," I said to Maria as we sat across from each other in April 2019. Her dark brown hair was done in a professional and flattering style. She was wearing a modest, collared shirt and slacks. She was smiling, alert, and looked put together, as she had for the past five years I'd been treating her.

Maria had been in sustained remission of her alcohol use disorder in all the time I had known her. She came to me already in recovery, achieved by attending Alcoholics Anonymous and working with her AA sponsor. She saw me occasionally to check in and refill her medications. I'm pretty sure I learned more from her than she ever did from me. One thing she taught me was that telling the truth was fundamental to her recovery.

Growing up she'd learned the opposite. Her mom drank, including blackout drinking and driving while Maria was in the car. Her dad left the family for several years for a place no one was allowed to name and which even now she'd rather not disclose out of respect for his privacy. It was left to her to take care of her younger siblings while pretending to the outside world that everything was fine at home. When Maria's own alcohol addiction began in her mid-twenties, she was already well practiced at shuffling through different versions of reality.

To illustrate the importance of honesty in her new sober life, she told me this story.

"I came home from work to find an Amazon package waiting for Mario."

Mario is Maria's younger brother. She and her husband, Diego, have been living with Mario as a way to support each other and save on rent in Silicon Valley's high-end real estate market.

"I decided to open it even though it wasn't addressed to me.

A part of me knew I shouldn't. When I'd opened his packages before, he got really mad. But I knew I could use the same excuse I'd used last time: that I mistook his name for mine, since they're so similar. I told myself I deserved one small pleasure after a long hard day of work. I don't remember now what was inside.

"After I opened the package, I resealed it and left it with the rest of the mail. To tell you the truth, I forgot about it. Mario came home a few hours later and immediately accused me of opening it. I lied and said I hadn't. He asked me again and I lied again. He kept saying, 'It looks like someone opened it.' I kept saying, 'It wasn't me.' Then he was really pissed and took his mail and his package and went into his room and slammed the door.

"I slept poorly that night. The next morning, I knew what I had to do. I walked into the kitchen where Mario and Diego were eating breakfast and said, 'Mario, I did open your package. I knew it was yours, but I opened it anyway. Then I tried to cover it up. Then I lied about it. I am so sorry. Please forgive me.'"

"Tell me why honesty is such an important part of your recovery," I said.

"I would never have admitted the truth back when I was drinking. Back then, I lied about everything and never took responsibility for the things I did. There were so many lies, and half of them didn't even make sense."

Maria's husband, Diego, once told me about how Maria used to hide in the bathroom to drink, turning on the shower so Diego wouldn't hear the sound of beer bottles opening, not realizing he could hear the clang of the bottle opener when she removed it from its hiding place behind the bathroom door. He described how she used to drink a six-pack in one sitting, then replace the beer with water and glue the tops back on. "Did she really think that I wouldn't be able to smell the glue, or taste the difference between water and booze?"

Maria said, "I lied to cover up my drinking, but I lied about other stuff too. Stuff that didn't even matter: where I was going, when I'd be back, why I was late, what I ate for breakfast."

Maria had developed the Lying Habit. What started out as a way to cover up her mother's drinking and her father's absence, and eventually her own addiction, turned into lying for its own sake.

The Lying Habit is remarkably easy to fall into. We all engage in regular lying, most of the time without realizing it. Our lies are so small and imperceptible that we convince ourselves we're telling the truth. Or that it doesn't matter, even if we know we're lying.

"When I told Mario the truth that day, even though I knew he'd be pissed, I knew something had really changed in me, in my life. I knew I was committed to living life in a different way, a better way. I was done with all those little lies filling up in the back of my mind and making me feel guilty and afraid . . . guilty for lying and afraid that someone would find out. I realized that as long as I'm telling the truth, I don't have to worry about any of that. I'm free. After I told my brother the truth about the package, it was a stepping-stone to our relationship getting closer. I went back upstairs after that and I felt really good."

Radical honesty—telling the truth about things large and small, especially when doing so exposes our foibles and entails consequences—is essential not just to recovery from addiction but for all of us trying to live a more balanced life in our reward-saturated ecosystem. It works on many levels.

First, radical honesty promotes awareness of our actions. Second, it fosters intimate human connections. Third, it leads to a truthful autobiography, which holds us accountable not just to our present but also to our future selves. Further, telling the truth is contagious, and might even prevent the development of future addiction.

Awareness

Earlier I described the Greek myth of Odysseus to illustrate physical self-binding. There's a little-known epilogue to this myth that is relevant here.

You'll remember that Odysseus asked his crew to tie him to the mast of his sailing ship to avoid the lure of the Sirens. But if you think about it, he could simply have put beeswax in his ears like he commanded the rest of his crew to do and saved himself a lot of grief. Odysseus wasn't a glutton for punishment. The Sirens could be killed only if whoever heard them could live to tell the story afterward. Odysseus vanquished the Sirens by narrating his near-death voyage after the fact. The slaying was in the telling.

The Odysseus myth highlights a key feature of behavior

change: Recounting our experiences gives us mastery over them. Whether in the context of psychotherapy, talking to an AA sponsor, confessing to a priest, confiding in a friend, or writing in a journal, our honest disclosure brings our behavior into relief, allowing us in some cases to see it for the first time. This is especially true for behaviors that involve a level of automaticity outside of conscious awareness.

When I was compulsively reading romance novels, I was only partially aware of doing so. That is to say, I was aware of the behavior at the same time I was not aware of it. This is a well-recognized phenomenon in addiction, a kind of half-conscious state akin to a waking dream, often referred to as *denial*.

Denial is likely mediated by a disconnect between the reward pathway part of our brain and the higher cortical brain regions that allow us to narrate the events of our lives, appreciate consequences, and plan for the future. Many forms of addiction treatment involve strengthening and renewing connections between these parts of the brain.

Neuroscientist Christian Ruff and his colleagues have studied the neurobiological mechanisms of honesty. In one experiment, they invited participants (145 total) to play a game in which they rolled dice for money using a computer interface. Before each roll, a computer screen indicated which outcomes would yield the monetary payoff, up to 90 Swiss francs (about 100 US dollars).

Unlike gambling in a casino, participants could lie about the results of the die roll to increase their winnings. The researchers were able to determine the degree of cheating by comparing the mean percentage of reported successful die rolls against the 50 percent benchmark implied by fully honest reporting. Not surprisingly, participants lied frequently. Compared with the 50 percent honesty benchmark, participants reported that 68 percent of their die rolls had the desired outcome.

Then the researchers used electricity to enhance neuronal excitability in the participants' prefrontal brain cortices, using a tool called transcranial direct current stimulation (tDCS). The prefrontal cortex is the frontmost part of our brain, just behind the forehead, and is involved in decision-making, emotion regulation, and future planning, among many other complex processes. It's also a key area involved in storytelling.

The researchers found that lying went down by half when neural excitability in the prefrontal cortex went up. In addition, the increase in honesty "could not be explained by changes in material self-interest or moral beliefs and was dissociated from participants' impulsivity, willingness to take risks, and mood."

They concluded that honesty can be strengthened by stimulating the prefrontal cortex, consistent with the idea that the "human brain has evolved mechanisms dedicated to control complex social behaviors."

This experiment led me to wonder if practicing honesty can stimulate prefrontal cortical activation. I e-mailed Christian Ruff in Switzerland to ask what he thought of this idea.

"If stimulating the prefrontal cortex causes people to be more honest, is it also possible that being more honest stimulates the prefrontal cortex? Might the practice of telling the truth strengthen activity and excitability in the parts of the brain we use for future planning, emotion regulation, and delayed gratification?" I asked.

He responded, "Your question makes sense. I have no definitive answer to it, but I share your intuition that a dedicated neural process (like the prefrontal process involved in honesty) should be strengthened by repeated use. This is what happens during most types of learning, according to Donald Hebb's old mantra, 'what fires together wires together.'"

I liked his answer because it implied that practicing radical honesty might strengthen dedicated neural circuits the same way that learning a second language, playing the piano, or mastering sudoku strengthens other circuits.

Consistent with the lived experience of people in recovery, truth-telling may change the brain, allowing us to be more aware of our pleasure-pain balance and the mental processes driving compulsive overconsumption, and thereby change our behavior.

My own dawning awareness of my problem with romance novels occurred in 2011 when I was teaching a group of San Mateo psychiatry residents how to talk to patients about addictive behaviors. The irony is not lost on me.

I was in a first-floor classroom of San Mateo Medical Center, giving a talk to nine psychiatry residents about how to have the often difficult conversations with patients about drug and alcohol use. I stopped partway through the lecture to invite the students to engage in a learning exercise: "Pair off with a partner to discuss a habit you want to change, and discuss some steps you might take to make that change."

Common examples of what students talk about in this exercise include "I want to exercise more" or "I want to eat less sugar." In other words, safer topics. Serious addictions, if they have them, don't usually get mentioned. Nonetheless, by talking about any behavior they're not happy with and want to change, students gain insight into what it might be like for patients to have these conversations with them as health care providers. There's also always the chance they'll discover something about themselves in the process.

I realized that with an odd number of students, I would have to partner with one student. I paired up with a soft-spoken, thoughtful young man who had been listening attentively throughout the lecture. I took the role of the patient so he could practice his skills. Then we would switch.

He asked me about a behavior I wanted to change. His gentle manner invited disclosure. To my surprise, I began to tell him an anodyne version of my late-night novel reading. I did not specify what I was reading or the extent of the problem.

I said, "I stay up too late at night reading, and it's interfering with my sleep. I'd like to change that."

As soon as I said it, I knew it was true, both that I was staying up too late reading and that I wanted to change the behavior. Until that moment, though, I wasn't really aware of either of those things.

"Why do you want to make that change?" he asked, using a standard question from motivational interviewing, a counseling approach developed by clinical psychologists William R.

Miller and Stephen Rollnick to explore internal motivations and resolve ambivalence.

"It's interfering with my ability to be as effective as I'd like to be at work and with my kids," I said.

He nodded. "Those sound like good reasons."

He was right. Those were good reasons. In saying them aloud, I realized for the first time how much my behavior was negatively impacting my life and the people I care about.

He then asked, "What would you be giving up if you stopped that behavior?"

"I'd be giving up the pleasure I get from reading. I love the escape," I answered right away. "But that feeling is not as important to me as my family and my work."

Again, by saying it out loud, I realized it was true: I value my family and my work above my own pleasure, and in order to live according to my values, I needed to stop the compulsive, escapist reading.

"What is one step you can take to change that behavior?"

"I can get rid of my electronic reader. Easy access to cheap reads fuels my late-night reading."

"Sounds like a good idea," he said, and smiled. We were done with me being the patient.

The next day, I kept thinking about our conversation. I decided to take a break from romance novels for the next month. The first thing I did was get rid of my e-reader. For the first two weeks, I experienced low-threshold withdrawal, including anxiety and insomnia, especially at night just before going to bed, a time when I would usually read stories. I had lost the art of falling asleep on my own.

At the end of the month, I was feeling better and gave myself permission to read romance again, planning to read in more moderation.

Instead, I binged on erotica, staying up late two nights in a row and feeling exhausted as a result. But now I saw my behavior for what it was—a compulsive, self-destructive pattern—which took the fun out of it. I felt a growing resolve to stop the behavior for good. My waking dream was coming to an end.

Honesty Promotes Intimate Human Connections

Telling the truth draws people in, especially when we're willing to expose our own vulnerabilities. This is counterintuitive because we assume that unmasking the less desirable aspects of ourselves will drive people away. It logically makes sense that people would distance themselves when they learn about our character flaws and transgressions.

In fact, the opposite happens. People come closer. They see in our brokenness their own vulnerability and humanity. They are reassured that they are not alone in their doubts, fears, and weaknesses.

Jacob and I met off and on over the months and years that followed his relapse to compulsive masturbation. In that time he continued to abstain from his addictive behaviors. Practicing radical honesty, especially with his wife, was the foundation of his ongoing recovery. At one of our visits he shared a story with me, something that happened shortly after he and his wife moved back in together.

She was sorting out the bathroom, a day after moving back into their shared home, when she noted one of the shower-curtain rings was missing. She asked Jacob if he knew what had happened to it.

"I freeze," Jacob told me. "I know perfectly what happen with the shower-curtain ring, but I don't want to tell her. I have many good reasons. It was long time ago. She will only be upset if I tell her. It is so good between us now. This will mess it up."

But then he reminded himself of how corrosive his lying and sneaking around had been for their relationship. He had promised her, before she moved back in, that he would be honest with her no matter what.

"So I say, 'I use it to build one of my machines, almost a year ago now, after you left. It is nothing recent. But I promise I will be honest with you, so I am telling you.'"

"What did she do?" I asked.

"I think she will tell me it is over and she is leaving again. But instead she does not yell at me. She does not leave me. She put her hand on my shoulder and she say, 'Thank you for telling me the truth.' And then she hug me."

Intimacy is its own source of dopamine. Oxytocin, a hormone much involved with falling in love, mother-child bonding, and lifetime pair bonding of sexual mates, binds to receptors on the dopamine-secreting neurons in the brain's reward pathway and enhances the firing of the reward-circuit tract. In other words, oxytocin leads to an increase in brain dopamine, a recent finding by Stanford neuroscientists Lin Hung, Rob Malenka, and their colleagues.

After his honest disclosure to his wife, followed by her expression of warmth and empathy, Jacob probably experienced a spike in oxytocin and dopamine in his reward pathway, encouraging him to do it again.

While truth-telling promotes human attachment, compulsive overconsumption of high-dopamine goods is the antithesis of human attachment. Consuming leads to isolation and indifference, as the drug comes to replace the reward obtained from being in relationship with others.

Experiments show that a free rat will instinctively work to free another rat trapped inside a plastic bottle. But once that free rat has been allowed to self-administer heroin, it is no longer interested in helping out the caged rat, presumably too caught up in an opioid haze to care about a fellow member of its species.

———

Any behavior that leads to an increase in dopamine has the potential to be exploited. What I'm referring to is a kind of "disclosure porn" that has become prevalent in modern culture, where revealing intimate aspects of our lives becomes a way to manipulate others for a certain type of selfish gratification rather than to foster intimacy through a moment of shared humanity.

At a medical conference on addiction in 2018, I sat next to a man who said he was in long-term recovery from addiction. He was there to tell his recovery story to the audience. Just before he went up on stage, he turned to me and said, "Get ready to cry." I was put off by the comment. It bothered me that he anticipated how I would react to his story.

He indeed told a harrowing story of addiction and recovery, but I was not moved to tears, which surprised me because I am usually deeply affected by stories of suffering and redemption. In this case, his story seemed untrue for all that it may have been factually correct. The words he spoke didn't match the emotions behind them. Instead of feeling that he was granting us privileged access to a painful time in his life, it felt like he was grandstanding and manipulating. Maybe it was just a matter of his having told it so many times before. In repetition, it may have grown stale. Whatever the reason, it didn't lift me.

There is a well-known phenomenon in AA called "drunkalogues," referring to tales of intoxicated exploits that are shared to entertain and show off rather than teach and learn. Drunkalogues tend to trigger craving rather than promote recovery. The line between honest self-disclosure and a manipulative drunkalogue is a fine one, including subtle differences in content, tone, cadence, and affect, but you know it when you see it.

I hope my disclosures here, my own and those my patients have given me permission to share, never stray to the wrong side of that line.

Truthful Autobiographies Create Accountability

Single, simple truths about our day-to-day lives are like links in a chain that translate into truthful autobiographical narratives. Autobiographical narratives are an essential measure of lived time. The stories we narrate about our lives not only serve as a measure of our past but can also shape future behavior.

In more than twenty years as a psychiatrist listening to tens of thousands of patient stories, I have become convinced that the *way* we tell our personal stories is a marker and predictor of mental health.

Patients who tell stories in which they are frequently the victim, seldom bearing responsibility for bad outcomes, are often unwell and remain unwell. They are too busy blaming others to get down to the business of their own recovery. By contrast, when my patients start telling stories that accurately portray their responsibility, I know they're getting better.

The victim narrative reflects a wider societal trend in which we're all prone to seeing ourselves as the victims of circumstance and deserving of compensation or reward for our suffering. Even when people have been victimized, if the narrative never moves beyond victimhood, it's difficult for healing to occur.

One of the jobs of good psychotherapy is to help people tell healing stories. If autobiographical narrative is a river, psychotherapy is the means by which that river is mapped and in some cases rerouted.

Healing stories adhere closely to real-life events. Seeking and finding the truth, or the closest approximation possible with the data at hand, affords us the opportunity for real insight and understanding, which in turn allows us to make informed choices.

As I have alluded to before, the modern practice of psychotherapy sometimes falls short of that lofty goal. We as mental-health care providers have become so caught up in the practice of empathy that we've lost sight of the fact that empathy without accountability is a shortsighted attempt to relieve suffering. If the therapist and patient re-create a story in which the patient is a perpetual victim of forces beyond their control, chances are good that the patient will continue to be victimized.

But if the therapist can help the patient take responsibility if not for the event itself, then for how they react to it in the here and now, that patient is empowered to move forward with their life.

I have been deeply impressed with AA philosophy and teachings on this point. One of the preeminent AA mottos, often printed in bold type on its brochures, is, "I am responsible."

In addition to responsibility, Alcoholics Anonymous emphasizes "rigorous honesty" as a central precept of its philosophy, and these ideas go together. The fourth of AA's 12 Steps requires members to take a "searching and fearless moral inventory," in which the individual considers his or her character defects and how they have contributed to a problem. The fifth step is the "confession step." This is where AA members "admit to God, to ourselves, and to another human being the exact nature of our wrong." This straightforward, practical, and systematic approach can have a powerful and transformative impact.

I personally experienced this in my thirties during my psychiatry residency training at Stanford.

My psychotherapy supervisor and mentor, the fedora-wearing one I mentioned at the very beginning, suggested I try the 12 Steps as a way to work through my resentments toward my mother. He realized long before I did that I was clinging to my anger in a ruminative and addictive way. I had spent years prior in psychotherapy trying to figure out my relationship with her, the effect of which seemed only to fuel my anger toward her for not being the mother I wanted her to be and the mother I thought I needed.

Through an act of generous self-disclosure, my supervisor shared with me that he was in decades-long recovery from an alcohol addiction, and that AA and the 12 Steps had helped him get there. Although my problem was not addiction per se, he had an instinctive sense that the 12 Steps would help me, and he agreed to walk me through it.

I worked the steps with him, and the experience was indeed transformative, especially Steps 4 and 5. For the first time in my life, rather than focusing on the ways I perceived my mother had failed me, I considered what I had contributed to our strained relationship. I concentrated on recent interactions rather than childhood events, as my responsibility during childhood was less.

At first it was difficult for me to see any ways I had contributed to the problem. I truly saw myself as the helpless victim in all regards. I was fixated on her reluctance to visit me in my home or cultivate a relationship with my husband and children, in contrast to her closer relationship with my siblings and their children. I resented what I perceived as her inability to accept me for who I am, and my sense that she wanted me to be someone different—someone warmer, more pliable, more self-effacing, less self-reliant, more fun.

But then I began engaging in the painful process of writing down . . . yes, writing down on paper and thereby making it very real indeed, my character defects and the ways those had contributed to our strained relationship. As Aeschylus said, "We must suffer, suffer into truth."

The truth is, I am anxious and fearful, although few would guess those things about me. I maintain a rigid schedule, a predictable routine, and a slavish adherence to my to-do list, as a way to manage my anxiety. This means that others are often forced to bend to my will and the exigencies of my goals.

Motherhood, although the most rewarding experience of

my life, has also been the most anxiety-provoking. Hence my defenses and ways of coping reached new heights when my children were little. Looking back, I realized it couldn't have been pleasant for anyone visiting our home during that time, including my own mother. I kept a tight grip on the running of our household and became acutely anxious when I perceived things to be out of order. I worked relentlessly, taking little or no time for myself, for friends and family, or for recreation. Indeed, I wasn't much fun in those days except, I hope, with my children.

As for my resentment toward my mother for wanting me to be different than I was, I realized with sudden and shocking clarity that I was guilty of the same thing toward her. I refused to accept her for who she was, wanting her instead to be some kind of Mother Teresa who would descend upon our home and care for all of us, including my husband and children, in just the way we needed to be cared for.

By demanding that she live up to some idealized vision of what I thought a mother and a grandmother should be, I was able to see only her flaws and none of her good qualities, of which she has many. She is a gifted artist. She is charming. She can be funny and zany. She has a kind heart and a giving nature as long as she doesn't feel judged or abandoned.

After working the steps, I was able to see the truth of these things more clearly, and with that, my resentment lifted. I was freed from the heavy burden of my anger toward my mother. What a relief!

My own healing contributed to an improvement in my relationship with her. I was less demanding, more forgiving, and less judgmental toward her. I also became aware of the many positive things resulting from our friction, namely, that I am resilient and self-reliant in ways I might not have been had she and I been more compatible.

I continue to try to practice that kind of truth-telling in all my relationships now. I'm not always successful, and instinctively want to pin the blame on others. But if I'm disciplined and diligent, I realize I too am responsible. When I'm able to get to that place and recount the real version to myself and others, I experience a feeling of rightness and fairness that gives the world the order I crave.

A truthful autobiographical narrative further allows us to be more authentic, spontaneous, and free in the moment.

The psychoanalyst Donald Winnicott introduced the concept of "the false self" in the 1960s. According to Winnicott, the false self is a self-constructed persona in defense against intolerable external demands and stressors. Winnicott postulated that the creation of the false self can lead to feelings of profound emptiness. No there there.

Social media has contributed to the problem of the false self by making it far easier for us, and even encouraging us, to curate narratives of our lives that are far from reality.

In his online life, my patient Tony, a young man in his twenties, ran every morning to take in the sunrise, spent the day engaged in constructive and ambitious artistic endeavors, and was the recipient of numerous awards. In his real life, he could barely get out of bed, compulsively looked at pornography online, struggled to find gainful employment, and was isolated, depressed, and suicidal. Little of his real day-to-day life was evident on his Facebook page.

When our lived experience diverges from our projected image, we are prone to feel detached and unreal, as fake as the false images we've created. Psychiatrists call this feeling *derealization* and *depersonalization*. It's a terrifying feeling, which commonly contributes to thoughts of suicide. After all, if we don't feel real, ending our lives feels inconsequential.

The antidote to the false self is the authentic self. Radical honesty is a way to get there. It tethers us to our existence and makes us feel real in the world. It also lessens the cognitive load required to maintain all those lies, freeing up mental energy to live more spontaneously in the moment.

When we're no longer working to present a false self, we're more open to ourselves and others. As the psychiatrist Mark Epstein wrote in his book *Going on Being* about his own journey toward authenticity, "No longer endeavoring to manage my environment, I began to feel invigorated, to find a balance, to permit a feeling of connection with the spontaneity of the natural world and with my own inner nature."

Truth-telling Is Contagious . . . and So Is Lying

In 2013, my patient Maria was at the height of her drinking problem. She was frequently presenting to the local emergency rooms with a blood alcohol level four times the legal limit. Diego, her husband, had assumed the bulk of caring for her.

Meanwhile, he was struggling with his own addiction to food. At five feet one, he weighed 336 pounds. It was only when Maria stopped drinking that Diego was motivated to tackle his food addiction.

"Seeing Maria get into recovery," he said, "motivated me to make changes to my own life. When Maria was drinking, I got away with a lot. I knew I was headed to a bad place. I didn't feel safe in my own body. But it was her getting sober that got me active. I could tell she was headed to a good place, and I didn't want to be left behind.

"So I got a Fitbit. I started going to the gym. I started calorie counting . . . just counting the calories made me realize how much I was eating. Then I started the keto diet and intermittent fasting. I wouldn't let myself eat late at night, or in the morning until I had worked out. I ran. I weight lifted. I realized hunger is a notification I can ignore. This year [2019] I weigh 195 pounds. I have a normal blood pressure for the first time in a long time."

In my clinical practice, I often see one member of a family get into recovery from addiction, followed quickly by another member of the family doing the same. I've seen husbands who stop drinking followed by wives who stop having affairs. I've seen parents who stop smoking pot followed by children who do the same.

I've mentioned the Stanford marshmallow experiment of 1968, in which children between the ages of three and six were studied for their ability to delay gratification. They were left alone in an empty room with a marshmallow on a plate and were told if they could go a full fifteen minutes without eating the marshmallow, they would get that marshmallow and a second one as well. They would get double the reward if they could just wait for it.

In 2012, researchers at the University of Rochester altered the 1968 Stanford marshmallow experiment in one crucial way. One group of children experienced a broken promise before the marshmallow test was conducted: The researchers left the room and said they would return when the child rang the bell, but then didn't. The other group of children were told the same, but when they rang the bell, the researcher returned.

The children in the latter group, where the researcher came back, were willing to wait up to four times longer (twelve minutes) for a second marshmallow than the children in the broken-promise group.

———

How can we understand why Maria's getting into recovery from her alcohol addiction inspired Diego to tackle his food problem; or why when adults keep their promises to children, those children are better able to regulate their impulses?

The way I understand this is by differentiating what I call the *plenty* versus the *scarcity mindset*. Truth-telling engenders a plenty mindset. Lying engenders a scarcity mindset. I'll explain.

When the people around us are reliable and tell us the truth, including keeping promises they've made to us, we feel more confident about the world and our own future in it. We feel we can rely not just on them but also on the world to be an orderly, predictable, safe kind of a place. Even in the midst of scarcity, we feel confident that things will turn out okay. This is a plenty mindset.

When the people around us lie and don't keep their promises, we feel less confident about the future. The world becomes a dangerous place that can't be relied upon to be orderly, predictable, or safe. We go into competitive survival mode and favor short-term gains over long-term ones, independent of actual material wealth. This is a scarcity mindset.

An experiment by the neuroscientist Warren Bickel and his colleagues looked at the impact on study participants' tendency to delay gratification for a monetary reward after having read a narrative passage that projected a state of plenty versus a state of scarcity.

The plenty narrative read like this: "At your job you have just been promoted. You will have the opportunity to move to a part of the country you always wanted to live in OR you may choose to stay where you are. Either way, the company gives you a large amount of money to cover moving expenses and tells you to keep what you don't spend. You will be making 100 percent more than you previously were."

The scarcity narrative read like this: "You have just been

fired from your job. You will now have to move in with a relative who lives in a part of the country you dislike, and you will have to spend all of your savings to move there. You do not qualify for unemployment, so you will not be making any income until you find another job."

The researchers found, not surprisingly, that participants who read the scarcity narrative were less willing to wait for a distant future payoff and more likely to want a reward now. Those who read the plenty narrative were more willing to wait for their reward.

It makes intuitive sense that when resources are scarce, people are more invested in immediate gains, and are less confident that those rewards will still be forthcoming in some distant future.

The question is, why do so many of us living in rich nations with abundant material resources nonetheless operate in our daily lives with a scarcity mindset?

As we have seen, having too much material wealth can be as bad as having too little. Dopamine overload impairs our ability to delay gratification. Social media exaggeration and "post-truth" politics (let's call it what it is, lying) amplify our sense of scarcity. The result is that even amidst plenty, we feel impoverished.

Just as it is possible to have a scarcity mindset amidst plenty, it is also possible to have a plenty mindset amidst scarcity. The feeling of plenty comes from a source beyond the material world. Believing in or working toward something outside ourselves, and fostering a life rich in human connectedness and meaning, can function as social glue by giving us a plenty mindset even in the midst of abject poverty. Finding connect-edness and meaning requires radical honesty.

Truth-telling as Prevention

"Let me first explain my role," I said to Drake, a physician I'd been asked to assess by our professional well-being com-mittee.

"I'm here to determine whether you might have a mental illness that adversely impacts your ability to practice medi-cine, and whether any reasonable accommodations are neces-sary for you to do your job. But I hope you will also see me as a resource beyond today's evaluation, should you need men-tal health treatment or emotional support more broadly."

"Thanks for that," he said, looking relaxed.

"I understand you got a DUI?"

DUI, or driving under the influence, is a legal infraction for operating a vehicle while intoxicated. For drivers twenty-one years or older in the United States, driving with a blood alcohol concentration (BAC) of 0.08 percent or higher is illegal.

"Yes. More than ten years ago, when I was in medical school."

"Hmm. I'm confused. Why are you seeing me now? Typi-cally, I'm asked to evaluate physicians in practice right after they get a DUI."

"I'm new on faculty here. I reported the DUI on my

application form. I guess they [the well-being committee] just wanted to make sure everything's okay."

"I guess that makes sense," I said. "Well, tell me your story."

In 2007, Drake was in his first semester of his first year of medical school. He'd driven out to the Northeast from California, trading the sunbaked grasslands of the Pacific Coast for the color-drenched rolling hills of New England in all their autumnal glory.

He had decided on medicine belatedly, some time after completing his undergraduate studies in California, where he'd effectively majored in surfing and spent one semester living in the woods behind campus, "writing bad poetry."

After the first exam, some of his med school classmates threw a party at their house in the country. The plan was for a friend to drive, but at the last minute, the friend had car trouble, so Drake ended up driving.

"I remember it was a beautiful early fall day in September. The house was down a country road, not far from where I was living."

The party turned out to be more fun than Drake expected. It was the first time he'd let loose since coming to medical school. He started out drinking a couple of beers, then progressed to Johnnie Walker Blue Label. By 11:30 p.m., when the cops showed up because of a neighbor calling in the noise, Drake was drunk. So was his friend.

"My friend and I realized we were too drunk to drive. So

we stayed at the house. I slept. The cops and most of the other guests left. I found a couch and tried to sleep it off. At 2:30 a.m., I got up. I was still a little drunk, but I didn't feel impaired. It was a straight shot down one empty country road back to my house. Two to three miles tops. We went for it."

As soon as Drake and his friend pulled onto the country road, they saw a police car waiting on the side of the road. The police pulled up behind them and started following them, as if they'd been waiting for them all along. They came to an intersection where there was a light signal hanging from one wire. It was blowing and twisting in the wind.

"I thought it was flashing yellow my way, and red the other way, but it was hard to tell with it swinging like that. Also, I was nervous with the cop right behind me. I went through the intersection slowly, and nothing happened, so I figured I was right about the flashing yellow, and I kept going. Just one more intersection, and a left turn to my house. I took the turn, but forgot to put on my blinker, and that's when the cop pulled me over."

The police officer was young, about the same age as Drake.

"He seemed new to the job, almost like he felt bad for pulling me over but had to do it."

He gave Drake a roadside sobriety test and Breathalyzed him. He blew 0.10 percent, just over the legal limit. The officer took Drake to the station, where Drake filled out a bunch of paperwork and learned that his license was temporarily suspended for driving under the influence. Someone from the station drove him home.

"The next day, I remembered a rumor that a friend I'd grown up with had gotten a DUI during his residency in emergency medicine. He was someone I really respected. He'd been our class president. I gave him a call."

"'Whatever you do,' my friend said when I reached him, 'you cannot get a DUI on your record, especially as a doctor. Get a lawyer immediately and they'll find a way to get it down to a "wet reckless" or get it off completely. That's what I did.'"

Drake found a local lawyer and paid him $5,000 up front, money he took out of his student loans.

The lawyer said to him, "They're going to assign you a court date. Dress up. Look nice. The judge is going to call you up to the stand and ask you how you plead, and you're going to say '*Not guilty.*' That's it. That's all you have to do. Two words. '*Not guilty.*' We'll take it from there."

On the day of his hearing, Drake dressed up like he was told to. He lived a few blocks from the courthouse, and as he walked there, he got to thinking. He thought about his cousin in Nevada who'd been driving while intoxicated and collided head-on with an eighteen-year-old girl coming the other way. They both died. People who saw his cousin in a bar just beforehand said he was drinking like he wanted to die.

"At the courthouse, I saw a bunch of other men about my age. They looked, you know, less privileged than me. I was thinking they probably didn't have a lawyer like I did. I started to feel a little sleazy."

Once inside the courtroom, waiting to be called, Drake kept running the plan through his head, just like his lawyer told him: "*The judge is going to call you up to the stand and ask you how you plead, and you're going to say 'Not guilty.' That's it. That's all you have to do. Two words. 'Not guilty.'"*

The judge called Drake to the witness stand. Drake settled into the hard wooden chair just below and to the right of the judge's bench. He was asked to raise his right hand and promise to tell the truth. He promised.

He looked out at the people in the courtroom. He looked at the judge. The judge turned to him and said, "How do you plead?"

Drake knew what he was supposed to say. He planned to say it. Two words. *Not guilty.* The words were almost on his lips. So close.

"But then I got to thinking about this time when I was five years old and I asked my dad for ice cream and he said I'd have to wait till after lunch. I told him, 'I ate lunch. I went next door to Michael's house, and he gave me a hot dog.' But the truth was I never went to Michael's house. Michael and I weren't really friends, and my dad knew it. Well, my dad didn't waste any time. He picked up the phone right then and asked Michael, 'Did you give Drake a hot dog?' Then my dad sat me down, totally calm, and told me it was always worse to lie. He said lying was never worth the consequences. That moment made a big impression on me.

"All along I'd been planning to plead 'not guilty,' just like the lawyer told me. It wasn't like I made a different decision before I took the stand. But the moment the judge asked me, I couldn't say the words. I just couldn't say them. I knew I was guilty. I had been drinking and driving."

"Guilty," Drake said.

The judge pulled up in his chair as if waking up for the first time that morning. Slowly he turned his head. He squinted his eyes right at Drake, drilling into him. "Are you sure that's your final plea? Do you realize the consequences? Because you can't go back."

"I'll never forget the way he swiveled his head and looked at me," Drake said. "I thought that was kind of odd that he was asking me that. I wondered for a split second if I was making a mistake. Then I told him I was sure."

Drake called the lawyer afterward and told him what happened. "He was definitely surprised."

Drake's lawyer said, "I respect your honesty. I don't usually do this, but I'm going to send you your five thousand dollars back."

And the lawyer did. A full refund.

Drake spent the next year attending mandatory DUI classes. The classes were in remote places. Since he couldn't drive, he had to take the bus, which could end up taking hours at a time. At the mandatory meetings, he sat in a circle with people he normally wouldn't have been exposed to. "A lot different from the people I was with in medical school." The other people in the class as he recalls were mostly older white men with multiple DUIs.

After paying over $1,000 in fines and spending tens of hours in mandatory DUI classes, Drake got his driver's license back. Turns out that was only the beginning.

He finished medical school and applied for residency, reporting the DUI conviction on all his residency applications.

When he applied for his medical license, he had to do the same thing. And again when he applied for specialty board certification. At the end of all that, when he took a residency position in the San Francisco Bay Area, he learned that none of the DUI classes he took in Vermont counted in California, so he had to do them all over again.

"I'd work these long days and into the night, then rush from the hospital to get to these meetings by bus. If I was one minute late, I had to pay a fee. There was a point then when I wondered if I would have been better off lying. But now, looking back, I'm glad I told the truth.

"Both my parents had drinking problems when I was growing up. My dad still does. He can go for weeks at a time and not drink, but when he does, it's not good. My mom has been in recovery for ten years now, but she was drinking the whole time I was growing up, though I didn't know it and never saw her drunk. But even with their problems, my parents were good about making me feel like I could be open and honest with them.

"They always seemed to have love and pride in me, even when I misbehaved. They didn't indulge me. They never gave me money to pay my legal fees, for example, though they had some money. But at the same time, they never judged me. I think they created a comfortable and safe space growing up. That allowed me to be open and honest.

"Today, I myself rarely drink. I'm prone to doing things in excess, and I'm a risk-taker, so I definitely could have gone that route. But I think telling the truth at that one crucial moment in my life, when I got that DUI, may have put me on another path. Maybe being honest over the years has helped me be more comfortable with myself. I have no secrets."

———

Telling the truth and suffering accelerated consequences may have changed the trajectory of Drake's life. He seemed to think so. The searing respect for honesty instilled in him by his father at an early age seemed to have a bigger impact than even his considerable genetic load for addiction. Could radical honesty be a preventive measure?

Drake's experience does not account for how radical honesty might backfire in a corrupt and dysfunctional system, or how the privileges of his race and class in American society contributed to his ability to overcome the considerable repercussions. Had he been poor and/or a person of color, the outcome might have looked very different.

Nonetheless, his story has convinced me as a parent that I can and should emphasize honesty as a core value in raising my children.

———

My patients have taught me that honesty enhances awareness, creates more satisfying relationships, holds us accountable to a more authentic narrative, and strengthens our ability to delay gratification. It may even prevent the future development of addiction.

For me, honesty is a daily struggle. There's always a part of me that wants to embellish the story just the slightest bit, to make myself look better, or to make an excuse for bad behavior. Now I try hard to fight that urge.

Although difficult in practice, this handy little tool—telling the truth—is amazingly within our reach. Anyone can wake up on any given day and decide, "Today I won't lie about anything." And in doing so, not just change their individual lives for the better, but maybe even change the world.

CHAPTER 9

Prosocial Shame

When it comes to compulsive overconsumption, shame is an inherently tricky concept. It can be the vehicle for perpetuating the behavior as well as the impetus for stopping it. So how do we reconcile this paradox?

First, let's talk about what shame is.

The psychological literature today identifies shame as an emotion distinct from guilt. The thinking goes like this: Shame makes us feel bad about ourselves as people, whereas guilt makes us feel bad about our actions while preserving a positive sense of self. Shame is a maladaptive emotion. Guilt is an adaptive emotion.

My problem with the shame-guilt dichotomy is that experientially, shame and guilt are identical. Intellectually, I may be able to parse out self-loathing from "being a good person who did something wrong," but in that moment of feeling shame-guilt, a gut punch of an emotion, the feeling is identical: regret mixed with fear of punishment and the terror of abandonment. The regret is for having been found out and may or may not include regret for the behavior itself. The terror of abandonment, its own form of punishment, is especially potent. It is the terror of being cast out, shunned, no longer part of the herd.

Yet the shame-guilt dichotomy is tapping into something real. I believe the difference is not how we experience the emotion, but how others respond to our transgression.

If others respond by rejecting, condemning, or shunning us, we enter the cycle of what I call *destructive shame*. Destructive shame deepens the emotional experience of shame and sets us up to perpetuate the behavior that led to feeling shame in the first place. If others respond by holding us closer and providing clear guidance for redemption/recovery, we enter the cycle of *prosocial shame*. Prosocial shame mitigates the emotional experience of shame and helps us stop or reduce the shameful behavior.

With that in mind, let's start by talking about when shame goes wrong (i.e., destructive shame) as a prelude to talking about when shame goes right (i.e., prosocial shame).

Destructive Shame

One of my psychiatry colleagues once said to me, "If we don't like our patients, we can't help them."

When I first met Lori, I didn't like her.

She was all business, quick to tell me she was there only because her primary care doctor sent her, which by the way was totally unnecessary because she had never had any kind of addiction or other mental health problem and just needed me to say as much so she could go back to the "real doctor" to get her meds.

"I've had gastric bypass surgery," she said, as if this should be explanation enough for the dangerously high doses of prescription drugs she was taking. Like an old-fashioned schoolmarm, she talked as if lecturing her less than gifted pupil. "I used to weigh over two hundred pounds and now I don't. So of course I have a malabsorption syndrome from rerouting my intestine, which is why I need 120 milligrams of Lexapro just to get to the blood levels of the average person. You, Doctor, of all people should understand that."

Lexapro is an antidepressant that modulates the neurotransmitter serotonin. Average daily doses are 10–20 mg, making Lori's dose at least six times the normal. Antidepressants are not typically misused to get high, but I have seen such cases over the years. Although it's true that the Roux-en-Y surgery that Lori received to lose weight can lead to a problem absorbing food and medications, it would be very unusual to need doses that high. Something else was going on.

"Are you using any other medications or any other substances?"

"I take gabapentin and medical marijuana for pain. I take Ambien for sleep. Those are my medicines. I need them to treat my medical conditions. I don't know what's wrong with that."

"What medical conditions are you treating?" Of course I had read her chart and knew what it said, but I always like to hear patients' understanding of their medical diagnosis and treatment.

"I have depression and pain in my foot from an old injury."

"Okay. That makes sense. But the doses are high. I'm wondering if you've ever struggled in your life with taking more of a substance or medication than you planned on, or using food or drugs to cope with painful emotions."

She stiffened, her back straight, her hands clasped in her lap, her ankles tightly crossed. She looked as if she might pop up from her chair and run out of the room.

"I told you, Doctor, I don't have that problem." She pursed her lips, then looked away.

I sighed. "Let's switch gears," I said, hoping to salvage our rough start. "Why don't you tell me about your life, like a mini autobiography: where you were born, who raised you, what you were like as a kid, major life milestones, all the way up to the present day."

Once I know a patient's story—the forces that shaped them to create the person I see before me—animosity evaporates in the warmth of empathy. To truly understand someone is to care for them. Which is why I always teach my medical students and residents—who are eager to parse experience into discrete boxes like "history of present illness," "mental status exam," and "review of systems" as they have been taught to do—to focus instead on story. Story recaptures not just the patient's humanity but also our own.

Lori grew up in the 1970s on a farm in Wyoming, the youngest of three siblings raised by her parents. She remembered from an early age feeling that she was different.

"Something wasn't right with me. I didn't feel like I belonged. I felt awkward and out of place. I had a speech impediment, a lisp. I felt stupid all my life." Lori was obviously whip-smart, but our early self-conceptions loom large in our lives, crowding out all evidence to the contrary.

She remembered being afraid of her father. He was prone to anger. But the bigger threat in their home was the specter of a punitive God.

"Growing up, I knew a damning God. If you weren't perfect, you were going to hell." As a result, telling herself she was perfect, or at least more perfect than other people, became an important theme throughout her life.

Lori was an average student and an above-average athlete. She set the middle school record in the 100-meter hurdles and began to dream about the Olympics. But in her junior year of high school, she broke her ankle running hurdles. She needed surgery, and her nascent running career effectively ended.

"The only thing I was good at got taken away. That's when I started eating. We'd stop at McDonald's and I could eat two Big Macs. I was proud of that. By the time I got to college, I didn't care about my appearance anymore. My freshman year I weighed 125 pounds. By the time I graduated and went to med tech school, I weighed 180 pounds. I also started experimenting with drugs: alcohol, marijuana, pills . . . mainly Vicodin. But my drug of choice was always food."

The next fifteen years of Lori's life were marked by wandering. Town to town, job to job, boyfriend to boyfriend. As a medical technician, it was easy to get work in almost any town. The one constant in Lori's life was that she attended church every Sunday, no matter where she was living.

During this time, she used food, pills, alcohol, cannabis, whatever she could get to escape from herself. In a typical day, she would eat a bowl of ice cream for breakfast, snack through work, and take an Ambien as soon as she got home. For dinner she'd eat another bowl of ice cream, a Big Mac, a Supersize fries, and a Diet Coke, followed by two more Ambien and a "big square of cake" for dessert. Sometimes she took Ambien at the end of her shift, getting a jump start so she could be high by the time she got home.

"If I didn't let myself sleep after I took it [the Ambien], I'd get a high. Then I'd take two more two hours later, and I'd get higher. Euphoric. Almost as good as opioids."

She'd repeat this cycle or a similar one day after day. On her vacation days, she'd mix sleeping pills with cough medicine to get a high, or drink alcohol to intoxication and engage in risky sexual behavior. By the time Lori was in her mid-thirties, she was living alone in a town house in Iowa, spending her leisure time getting high and listening to American radio host Glenn Beck.

"I became convinced the end of the world was coming. Armageddon. Muslims. An Iranian invasion. I bought a bunch of gas in containers. I stored them in my extra bedroom. Then I put them on the patio under a tarp. I bought a .22 caliber rifle. Then I realized I could blow up, so I started filling my car with gas from the containers until it was all gone."

On some level, Lori knew she needed help, but she was terrified to ask for it. She was afraid that if she admitted she wasn't the "perfect Christian," people would recoil from her. She had on occasion hinted at her problems with fellow church members but came to understand through subtle messaging that there were certain types of problems congregants weren't supposed to share. At that point she weighed almost 250 pounds, felt a crushing depression, and began to wonder if she might be better off dead.

"Lori," I said, "when we look at the whole, whether food or cannabis or alcohol or prescription pills, one of the enduring problems seems to be compulsive, self-destructive overconsumption. Do you think that's fair?"

She looked at me and didn't say anything. Then she began to cry. When she was able to speak, she said, "I know it's true, but I don't want to believe it. I don't want to hear it. I have a job. I have a car. I go to church every Sunday. I thought having the gastric bypass surgery would fix everything. I thought losing the weight would change my life. Even when I lost the weight, I still wanted to die."

I suggested a number of different paths Lori might take to get better, including attending AA.

"I don't need that," she said without hesitation. "I've got my church."

A month later, Lori came back as scheduled.

"I met with the church elders."

"What happened?"

She looked away. "I was open in a way I'd never been before . . . except with you. I told them everything . . . or almost everything. I just put it all out there."

"And?"

"It was weird," she said. "They seemed . . . confused. Anxious. Like they didn't really know what to do with me. They told me to pray. They said they would pray for me. They also encouraged me not to discuss my problems with other members of the church. That's it."

"How was that for you?"

"At that moment I felt that damning, shaming God. I'm able to quote Scripture but I feel no connection to the loving God of Scripture. I can't live up to that expectation. I'm not that good. So I stopped going to church. I haven't been in a month. And do you know, no one has seemed to notice. No one called. No one contacted me. Not one person."

Lori was caught in the cycle of destructive shame. When she tried being honest with fellow church members, she was discouraged from sharing that part of her life, implicitly communicating that she would be rejected or further shamed if she were open about her struggles. She couldn't risk losing what little community she had. But keeping her behavior hidden also perpetuated her shame, further contributing to isolation, all of which fueled ongoing consumption.

Studies show that people who are actively involved in religious organizations on average have lower rates of drug and alcohol misuse. But when faith-based organizations end up on the wrong side of the shame equation, by shunning transgressors and/or encouraging a web of secrecy and lies, they contribute to the cycle of destructive shame.

Destructive shame looks like this: Overconsumption leads to shame, which leads to shunning by the group or lying to the group to avoid shunning, both of which result in further isolation, contributing to ongoing consumption as the cycle is perpetuated.

The antidote to destructive shame is *prosocial shame*. Let's see how that might work.

AA as a Model for Prosocial Shame

My mentor once told me about what motivated him to stop drinking alcohol. I've often recalled his story because it illustrates the twin-edged knife blade of shame.

Well into his forties he would drink secretly every night after his wife and children went to bed. He did it long after he promised his wife he had stopped. All the little lies he told to cover up his drinking, and the fact of the drinking itself, accumulated and weighed on his conscience, which in turn led him to drink more. He drank for shame.

One day his wife discovered his use. "The look of disappointment and betrayal in her eyes made me swear I would never drink again." The shame he felt in that moment, and his desire to regain the trust and approval of his wife, propelled him into his first serious attempt at recovery. He started attending Alcoholics Anonymous meetings. He identified the main benefit of Alcoholics Anonymous for him as a "de-shaming process."

He described it this way. "I realized I wasn't the only one. There were other people just like me. There were other doctors who were struggling with alcohol addiction. Knowing I had a place to go where I could be completely honest and still be accepted was incredibly important. It created the psychological space I needed to forgive myself and make changes. To move forward in my life."

Prosocial shame is predicated on the idea that shame is useful and important for thriving communities. Without shame, society would descend into chaos. Hence, feeling shame for transgressive behaviors is appropriate and good.

Prosocial shame is further predicated on the idea that we are all flawed, capable of making mistakes, and in need of forgiveness. The key to encouraging adherence to group norms, without casting out every person who strays, is to have a post-shame "to-do" list that provides specific steps for making amends. This is what AA does with its 12 Steps.

The prosocial shame cycle goes like this: Overconsumption leads to shame, which demands radical honesty and leads *not* to shunning, as we saw with destructive shame, but to acceptance and empathy, coupled with a set of required actions to make amends. The result is increased belonging and decreased consumption.

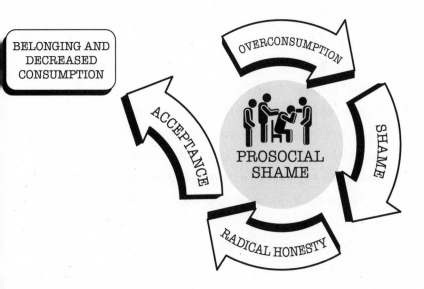

My patient Todd, a young surgeon in recovery from alcohol addiction, told me how AA "was the first safe place to express vulnerability." At his first AA meeting he cried so hard he wasn't able to say his name.

"Afterward, everyone came up, gave me their numbers, told me to call. It was that community I always wanted but never had. I could never have opened up like that with my rock-climbing friends or with other surgeons."

After five years in sustained recovery, Todd shared with me that the most important step of the 12 Steps for him was Step 10 ("Continued to take personal inventory and when we were wrong promptly admitted it").

"Every day, I check in with myself. *Okay, am I twisted? If yes, how can I change it? Do I need to make amends? How can I make amends?* For example, the other day, I was dealing with a resident who didn't give me the right information about a patient. I started to get frustrated. *Why isn't this being done?* When I feel that frustration, I tell myself: *Okay, Todd, stop. Think about this. This person has almost ten years less experience than you. They're probably scared. Instead of getting frustrated, how can you help them get what they need?* That's not something I would have done prior to getting into recovery.

"A couple of years ago," Todd told me, "about three years into my recovery, I was supervising a medical student who was just awful. I mean really bad. I would not let him take care of patients. When it came time for midterm feedback, I sat down with him and decided to be honest. I told him, 'You're not going to pass this rotation unless you make some big changes.'

"After my feedback. he decided to start over and really try to improve his performance. He was able to get better and he did end up passing the rotation. The thing is, in my drinking days, I wouldn't have been honest with him. I would have just let him go on and fail the rotation, or leave the problem for someone else to deal with."

A truthful self-inventory leads not only to a better understanding of our own shortcomings. It also allows us to more objectively appraise and respond to the shortcomings of others. When we're accountable to ourselves, we're able to hold others accountable. We can leverage shame without shaming.

The key here is accountability with compassion. These lessons apply to all of us, addicted or not, and translate to every type of relationship in our everyday lives.

———

Alcoholics Anonymous is a model organization for prosocial shame. Prosocial shame in AA leverages adherence to group norms. There is no shame about being an "alcoholic," consistent with the saying "AA is a no-shame zone;" but there is shame about the half-hearted pursuit of "sobriety." Patients have told me that the anticipated shame of having to admit to the group they've relapsed works as a major deterrent against relapse, and promotes further adherence to group norms.

Importantly, when AA members do relapse, the relapse itself is a club good. Behavioral economists refer to the rewards of belonging to a group as *club goods*. The more robust the club goods, the more likely the group will be able to maintain its current members and attract new members. The concept of club goods can be applied to any group of humans, from families to friendship groups to religious congregations.

As the behavioral economist Laurence Iannaccone has written in reference to club goods in faith-based organizations, "The pleasure I derive from Sunday service depends not just on my own inputs but also on the inputs of others: how many others attend, how warmly they greet me, how well they sing, how enthusiastically they read and pray." Club goods are strengthened by active participation in group activities and gatherings, and by adherence to group rules and norms.

The honest disclosure of a relapse to the AA fellowship augments club goods by creating the opportunity for other group members to experience empathy, altruism, and, let's face it, some degree of schadenfreude along the lines of "That could have happened to me and I'm sure glad it didn't," or "There but for the grace of God go I."

Club goods are threatened by free riders who attempt to benefit from the group without sufficient participation in that community, similar to the more colloquial terms *free-loaders* or *moochers*. When it comes to group rules and norms, free riders threaten club goods when they fail to adhere, lie about it, and/or make no effort to change their behavior. Their individual behavior is doing nothing to strengthen club goods, yet they're individually benefiting from having membership in the group—the bennies of belonging.

Iannaconne noted that it is difficult if not impossible to measure adherence to the group principles that create the club goods, especially when the demands involve personal habits and nontangible, subjective phenomena, such as truth-telling.

Iannaconne's *Theory of Sacrifice and Stigma* posits that one way to "measure" group participation is indirectly, by mandating stigmatizing behaviors that reduce participation in other contexts, and by demanding the sacrifice of the individual's resources to the exclusion of other activities. Thus are free riders ferreted out.

In particular, those behaviors that seem excessive, gratuitous, or even irrational in existing religious institutions, such as wearing certain hairstyles or certain clothing, abstaining from various foods or forms of modern technology, or refusing certain medical treatments, are rational when understood as a cost to the individual to reduce free riding within an organization.

You might think that religious organizations and other social groups that are more relaxed, with fewer rules and strictures, would attract a larger group of followers. Not so. "Stricter churches" achieve a larger following and are generally more successful than freewheeling ones because they ferret out free riders and offer more robust club goods.

Jacob joined the 12 Step group Sexaholics Anonymous (SA) early in his recovery process and stepped up his involvement each time he relapsed. The commitment was formidable. He attended a group meeting in person or by phone daily. He often made eight or more phone calls each day with fellow members.

AA and other 12 Step groups have been maligned as "cults" or organizations in which people trade their addiction to alcohol and/or drugs to an addiction to the group. These criticisms fail to appreciate that the strictness of the organization, its cultishness, may be the very source of its effectiveness.

Free riders in 12 Step groups can take many forms, but

among the most dangerous are those members who do not admit when they've relapsed, do not re-declare themselves as newcomers, and do not rework the steps. They deprive the group of the club good of prosocial shame, not to mention the sober social network crucial to recovery. To maintain club goods, AA must take strong and at times seemingly irrational measures against this type of free riding.

Joan was able to quit drinking through her participation in AA. She too went to regular meetings, had a sponsor, and herself sponsored others. She had been abstinent from alcohol in AA for four years and my patient for ten, so I was able to observe and appreciate all the positive changes AA had made in her life.

Joan had an incident in the early 2000s in which she used alcohol unwittingly. She was traveling in Italy, where she did not speak the language, and accidentally ordered and consumed a beverage that contained a very small percentage of alcohol, on par with nonalcoholic beers marketed and sold in the United States. It was only afterward that she realized what had happened, not because she felt altered but because she read the label.

When she returned from her trip and told her sponsor what had happened, her sponsor insisted she had relapsed and encouraged her to tell the group and reset her sobriety date. I was surprised that Joan's sponsor took such a rigid stance. After all, she consumed an amount of alcohol so negligible that most Americans do not consider such beverages to be "alcoholic." But Joan agreed, although she did so tearfully.

She has maintained her recovery and her participation in AA to this day.

Joan's sponsor's insistence that she reset her sobriety date seemed excessive to me at the time, but now I understand it as both guarding against a little bit of alcohol giving way to a lot of alcohol—the slippery slope—and "utility maximizing" for the greater good of the group. Joan's willingness to abide by a very stringent interpretation of relapse strengthened her ties to the group, which turned out to be positive in the long run for her as well.

Also, Joan herself pointed out, "Maybe there was a part of me that knew there was alcohol in the drink and wanted to use being in a foreign country as an excuse." In that sense the group functions as an extended conscience.

Of course, groupthink strategies can be used to nefarious ends. For example, when the cost of belonging exceeds club goods and members are harmed. NXIVM was a self-described Executive Success Program whose leaders were arrested and indicted in 2018 on federal sex trafficking and racketeering charges. Similarly, there are situations in which members of a group benefit, but they harm those outside the group, such as various entities today who use social media to spread falsehoods.

———

A few months after stopping church, Lori went to her first AA meeting. AA provided the supportive fellowship she was looking for but unable to find at her church. On December 20, 2014, Lori quit all substances and has maintained her recovery since.

"I can't tell you exactly what happened, or when," Lori said, looking back at her own recovery years later, which she credits to her participation in AA. "Hearing people's stories. The relief I felt letting go of my deepest, darkest secrets. Seeing the hope in newcomers' eyes. I was so isolated before. I remember just wanting to die. Lying awake at night whipping myself for all the things I'd done. In AA, I learned to accept myself and other people for who they are. Now I have real relationships with people. I belong. They know the real me."

Prosocial Shame and Parenting

As a parent who is worried about her children's well-being in a world flooded with dopamine, I've tried to incorporate the principles of prosocial shame into our family life.

First, we've established radical honesty as a core family value. I try hard, not always with success, to model radical honesty in my own behavior. Sometimes as parents we think that by hiding our mistakes and imperfections and only revealing our best selves, we'll teach our children what is right. But this can have the opposite effect, leading children to feel they must be perfect to be lovable.

Instead, if we are open and honest with our children about our struggles, we create a space for them to be open and honest about their own. As such, we must also be ready and willing to admit when we've been wrong in our interactions with them and with others. We must embrace our own shame and be willing to make amends.

About five years ago, when our kids were still in elementary and junior high school, I gave them each a chocolate bunny for Easter. Made of creamy milk chocolate, they were from a specialty chocolatier. My children ate a little of their bunnies and put the rest away in the pantry for later.

Over the following two weeks, I nibbled a little here and a little there at their chocolate bunnies, not enough, I thought, for anyone to notice. By the time my kids remembered their chocolate bunnies, I had whittled them down to almost nothing. Knowing my affinity for chocolate, they accused me first.

"It wasn't me," I said. The lie came naturally. I continued to lie over the next three days. They persisted in being skeptical that I was telling the truth, but then they began accusing each other. I knew I had to make it right. *How will I teach my children honesty if I'm not honest myself? And what a silly, stupid thing to lie about!* It took me three days to build up the courage to tell them the truth. I was so ashamed.

They were vindicated and horrified at learning the truth. Vindicated that their first guess had been right. Horrified that their own mother would lie to them. It was instructive for me and them on many levels.

I reminded myself, and signaled to them, how deeply flawed I am. I also modeled that when I make mistakes I can at least own my part. My kids forgave me and to this day love to tell the story about how I "stole" their chocolate and then "lied about it." Their teasing is my penance and I welcome it. Together we reaffirmed as a family that ours is one in which people will make mistakes but not be permanently condemned or cast out. We're learning and growing together.

Like my patient Todd, when we engage in an active and honest reappraisal of ourselves, we're more able and willing to give other people honest feedback, in the spirit of helping them understand their own strengths and shortcomings.

This type of radical honesty without shaming is also important to teach children their strengths and weaknesses.

When our elder daughter was five years old, she started piano lessons. I was raised in a musical family and looked forward to sharing music with my children. It turned out my daughter had no sense of rhythm and although not quite tone-deaf, came pretty close. Yet we both doggedly persisted with her daily practice, me sitting next to her, trying to be encouraging, while containing my horror at her utter lack of aptitude. The truth is neither of us enjoyed it.

About a year into her lessons, we were watching the movie *Happy Feet* about a penguin, Mumble, who has a big problem: He can't sing a single note, in a world where you need a heart song to attract a soul mate. Our daughter looked at me halfway through the movie and said, "Mom, am I like Mumble?"

I was gripped in the moment by parental self-doubt. *What*

do I say? Do I tell her the truth and risk damaging her self-esteem, or do I lie and try to use the deception to kindle a love of music?

I took a risk. "Yes," I said, "you're pretty much like Mumble."

A big smile broke out across my daughter's face, which I interpreted as the smile of validation. I knew then I had done the right thing.

In validating what she already knew to be true—her lack of musical ability—I encouraged her skills at accurate self-appraisal, skills she continues to demonstrate to this day. I also sent the message that we can't be good at everything, and it's important to know what you're good at and what you're not good at, so you can make wise decisions.

She decided to quit piano lessons after a year—to everyone's relief—and she enjoys music to this day, singing along with the radio completely off-key and not the least bit embarrassed by it.

Mutual honesty precludes shame and presages an intimacy explosion, a rush of emotional warmth that comes from feeling deeply connected to others when we're accepted despite our flaws. It is not our perfection but our willingness to work together to remedy our mistakes that creates the intimacy we crave.

This kind of intimacy explosion is almost certainly accompanied by the release of our brain's own endogenous dopamine. But unlike the rush of dopamine we get from cheap pleasures, the rush we get from true intimacy is adaptive, rejuvenating, and health-promoting.

Through sacrifice and stigma, my husband and I have attempted to strengthen our family's club goods.

Our kids were not allowed to have their own phone until they got to high school. This made them an oddity among their peers, especially in middle school. At first they begged and cajoled for a phone of their own, but after a while came to see this difference as a core part of their identity, along with our insistence that we bike instead of drive whenever possible, and spend time together as a family without devices.

I'm convinced that our kids' swim coach has a secret PhD in behavioral economics. He leverages sacrifice and stigma on a regular basis to strengthen club goods.

First, there's the prodigious time commitment, up to four hours a day of swim practice for kids in high school, and the covert shaming that happens when kids miss practice. There's recognition and rewards for high attendance (not unlike AA's token for thirty meetings in thirty days), including the opportunity to participate in travel meets. There's the strict guidelines on what to wear to meets: red swim T-shirts on Fridays, gray swim T-shirts on Saturday, team-logo apparel (caps, suits, goggles) only. This successfully distinguishes the kids on this team from the casual appearance of kids on other teams.

Many of these rules seem excessive and gratuitous, but when viewed through the lens of utility-maximizing principles to strengthen participation, reduce free riding, and augment club goods, they make sense. And kids flock to this team in particular, seeming to love the strictness, even as they complain about it.

We tend to think of shame as a negative, especially at a time when *shaming*—fat shaming, slut shaming, body shaming, and so on—is such a loaded word and is (rightly) associated with bullying. In our increasingly digital world, social media shaming and its correlate "cancel culture" have become a new form of shunning, a modern twist on the most destructive aspects of shame.

Even when no one else is pointing the finger at us, we're all too ready to point it at ourselves.

Social media propels our tendency toward self-shame by inviting so much invidious distinction. We're now comparing ourselves not just to our classmates, neighbors, and coworkers, but to the whole world, making it all too easy to convince ourselves that we should have done more, or gotten more, or just lived differently.

To deem our lives "successful," we now feel we must achieve the mythic heights of Steve Jobs and Mark Zuckerberg or, like the Theranos corporation's Elizabeth Holmes, a latter-day Icarus, go down in flames trying.

But the lived experience of my patients suggests that prosocial shame can have positive, healthy effects by smoothing some of narcissism's rougher edges, tying us more closely to our supportive social networks, and curbing our addictive tendencies.

Lessons of the Balance

We all desire a respite from the world—a break from the impossible standards we often set for ourselves and others. It's natural that we would seek a reprieve from our own relentless ruminations: *Why did I do that? Why can't I do this? Look what they did to me. How could I do that to them?*

So we're drawn to any of the pleasurable forms of escape that are now available to us: trendy cocktails, the echo chamber of social media, binge-watching reality shows, an evening of Internet porn, potato chips and fast food, immersive video games, second-rate vampire novels . . . The list really is endless. Addictive drugs and behaviors provide that respite but add to our problems in the long run.

What if, instead of seeking oblivion by escaping from the world, we turn toward it? What if instead of leaving the world behind, we immerse ourselves in it?

Muhammad, you'll remember, was my patient who tried various forms of self-binding to limit his cannabis consumption, only to find himself right back where he started, progressing from moderation to excessive consumption to addiction at an ever-faster cadence.

He went hiking at Point Reyes, a nature trail just north of San Francisco, in hopes of finding refuge in an activity that had previously given him pleasure, as he once again tried to get control of his cannabis consumption.

But every turn in the bend brought fresh memories of smoking weed—hiking trips in the past had almost always occurred in a state of semi-intoxication—and so, instead of being an escape, the hike turned into an agony of craving and a painful reminder of loss. He despaired of ever being able to wrestle his cannabis problem into submission.

Then he had his aha moment. At one particular vista point where he had explicit memories of smoking a joint with friends, he brought the camera up to his eye and pointed it to a nearby plant. He saw a bug on a leaf and focused the camera further, zooming in on the beetle's bright red carapace, striated anten-nae, and ferociously hairy legs. He was mesmerized.

His attention was snared by the creature in his crosshairs. He took a series of pictures, then changed his angle and took more. For the rest of the hiking trip, he stopped to take ex-tremely close-up pictures of beetles. As soon as he did so, his cravings for cannabis decreased.

"I had to force myself to be very still," he told me at one of our sessions in 2017. "I had to achieve a perfect stillness to take a good in-focus picture. That process grounded me, lit-erally, and centered me. I discovered a strange, surreal, and compelling world at the end of my camera that rivaled the world I escaped to with drugs. But this was better because no drugs were needed."

Many months later, I realized Muhammad's path to recovery was similar to my own.

I made a conscious decision to reimmerse myself in patient care, focusing on the aspects of my work that had always been rewarding: relationships with my patients over time, and immersion in narrative as a way to bring order to the world. In doing so, I was able to emerge from compulsive romance reading into a more rewarding and meaningful career. I was also more successful in my work, but my success was an unexpected byproduct, not the thing I was seeking.

I urge you to find a way to immerse yourself fully in the life that you've been given. To stop running from whatever you're trying to escape, and instead to stop, and turn, and face whatever it is.

Then I dare you to walk toward it. In this way, the world may reveal itself to you as something magical and awe-inspiring that does not require escape. Instead, the world may become something worth paying attention to.

The rewards of finding and maintaining balance are neither immediate nor permanent. They require patience and maintenance. We must be willing to move forward despite being uncertain of what lies ahead. We must have faith that actions today that seem to have no impact in the present moment are in fact accumulating in a positive direction, which will be revealed to us only at some unknown time in the future. Healthy practices happen day by day.

My patient Maria said to me, "Recovery is like that scene in

Harry Potter when Dumbledore walks down a darkened alley lighting lampposts along the way. Only when he gets to the end of the alley and stops to look back does he see the whole alley illuminated, the light of his progress."

Here we are at the end, but it could be just the beginning of a new way of approaching the hypermedicated, overstimulated, pleasure-saturated world of today. Practice the lessons of the balance, so that you too can look back at the light of your progress.

Lessons of the Balance

1. *The relentless pursuit of pleasure (and avoidance of pain) leads to pain.*

2. *Recovery begins with abstinence.*

3. *Abstinence resets the brain's reward pathway and with it our capacity to take joy in simpler pleasures.*

4. *Self-binding creates literal and metacognitive space between desire and consumption, a modern necessity in our dopamine-overloaded world.*

5. *Medications can restore homeostasis, but consider what we lose by medicating away our pain.*

6. *Pressing on the pain side resets our balance to the side of pleasure.*

7. *Beware of getting addicted to pain.*

8. *Radical honesty promotes awareness, enhances intimacy, and fosters a plenty mindset.*

9. *Prosocial shame affirms that we belong to the human tribe.*

10. *Instead of running away from the world, we can find escape by immersing ourselves in it.*

Author's Note

The intimate conversations and stories in this book are included with the interviewees' informed consent. In order to protect privacy, I've deleted and changed names and other demographic details even when participants were willing to include them unchanged. The process of obtaining consent included participants assenting to the following: "Someone who knows you well and reads your story here will likely recognize you even though I've changed your name. Are you okay with that?" And "If there are any details you don't want me to include, let me know and I'll leave them out."

Notes

2 "contemporary prophets that we ignore": Kent Dunnington, *Addiction and Virtue: Beyond the Models of Disease and Choice* (Downers Grove, IL: InterVarsity Press Academic, 2011). This is a wonderful theological and philosophical treatise on addiction and faith.

18 US opioid epidemic: Anna Lembke, *Drug Dealer, MD: How Doctors Were Duped, Patients Got Hooked, and Why It's So Hard to Stop*, 1st ed. (Baltimore: Johns Hopkins University Press, 2016). There are many excellent books on this topic, including *Pain Killer: An Empire of Deceit and the Origin of America's Opioid Epidemic*, by Barry Meier; *Dreamland: The True Tale of America's Opiate Epidemic*, by Sam Quinones; and *Dopesick: Dealers, Doctors and the Drug Company That Addicted America*, by Beth Macy. Each of these books, including my own, explores the origins of the opioid epidemic through a slightly different lens.

19 "tremendous expansion of the supply": ASPPH Task Force on Public Health Initiatives to Address the Opioid Crisis, *Bringing Science to Bear on Opioids: Report and Recommendations*, November 2019.

19 "repeated exposure to opioids": ASPPH Task Force on Public Health Initiatives to Address the Opioid Crisis, *Bringing Science to Bear on Opioids: Report and Recommendations*, November 2019.

19 Prohibition led to a sharp decrease: Wayne Hall, "What Are the Policy Lessons of National Alcohol Prohibition in the United States, 1920–1933?," *Addiction* 105, no. 7 (2010): 1164–73, https://doi.org/10.1111/j.1360-0443.2010.02926.x.

19 There were unintended consequences: Robert MacCoun, "Drugs and the Law: A Psychological Analysis of Drug Prohibition," *Psychological Bulletin* 113 (June 1, 1993): 497–512, https://doi.org/10.1037//0033-2909.113.3.497. There's considerable controversy and debate on the impact of prohibition, decriminalization, and legalization of psychoactive drugs. Rob MacCoun's work on this topic blends economics, psychology, and political philosophy for a deep dive.

20 **diagnosable alcohol addiction rose by 50 percent**: Bridget F. Grant, S. Patricia Chou, Tulshi D. Saha, Roger P. Pickering, Bradley T. Kerridge, W. June Ruan, Boji Huang, et al., "Prevalence of 12-Month Alcohol Use, High-Risk Drinking, and DSM-IV Alcohol Use Disorder in the United States, 2001–2002 to 2012–2013: Results from the National Epidemiologic Survey on Alcohol and Related Conditions," *JAMA Psychiatry* 74, no. 9 (September 1, 2017): 911–23, https://doi.org/10.1001/jamapsychiatry.2017.2161.

20 **Mental illness is a risk factor**: Anna Lembke, "Time to Abandon the Self-Medication Hypothesis in Patients with Psychiatric Disorders," *American Journal of Drug and Alcohol Abuse* 38, no. 6 (2012): 524–29, https://doi.org/10.3109/00952990.2012.694532.

20 **"limbic capitalism"**: David T. Courtwright, *The Age of Addiction: How Bad Habits Became Big Business* (Cambridge, MA: Belknap Press, 2019), https://doi.org/10.4159/9780674239241. This is a gripping and erudite look at the way that increased access to addictive goods and behaviors through time and across cultures has contributed to increased consumption.

20 **The cigarette-rolling machine**: Matthew Kohrman, Gan Quan, Liu Wennan, and Robert N. Proctor, eds., *Poisonous Pandas: Chinese Cigarette Manufacturing in Critical Historical Perspectives* (Stanford, CA: Stanford University Press, 2018).

21 **morphine addiction:** David T. Courtwright, "Addiction to Opium and Morphine," in *Dark Paradise: A History of Opiate Addiction in America* (Cambridge, MA: Harvard University Press, 2009), https://doi.org/10.2307/j.ctvk12rbo.7. This is another fantastic book by historian David Courtwright, tracing the origins of the opioid epidemic through history, including the late 1800s, when physicians routinely prescribed morphine to Victorian housewives, among others.

22 **automation of chip and fry:** National Potato Council, *Potato Statistical Yearbook 2016*, accessed April 18, 2020, https://www.nationalpotatocouncil.org/files/7014/6919/7938/NPCyearbook2016_-_FINAL.pdf.

22 Thai tomato coconut bisque: Annie Gasparro and Jessie Newman, "The New Science of Taste: 1,000 Banana Flavors," *Wall Street Journal*, October 31, 2014. Also see David T. Courtwright's *The Age of Addiction: How Bad Habits Became Big Business,* for an excellent, extended discussion on changes in the food industry.

29 leading global risks for mortality: Shanthi Mendis, Tim Armstrong, Douglas Bettcher, Francesco Branca, Jeremy Lauer, Cecile Mace, Vladimir Poznyak, Leanne Riley, Vera da Costa e Silva, and Gretchen Stevens, *Global Status Report on Noncommunicable Diseases 2014* (World Health Organization, 2014), https://apps.who.int/iris /bitstream/handle/10665/148114/9789241564854_eng.pdf.

29 now more people worldwide . . . who are obese: Marie Ng, Tom Fleming, Margaret Robinson, Blake Thomson, Nicholas Graetz, Christopher Margono, Erin C Mullany, et al., "Global, Regional, and National Prevalence of Overweight and Obesity in Children and Adults during 1980–2013: A Systematic Analysis for the Global Burden of Disease Study 2013," *Lancet* 384, no. 9945 (August 2014): 766–81, https://doi.org/10.1016/S0140-6736(14)60460-8.

29 Global deaths from addiction have risen: Hannah Ritchie and Max Roser, "Drug Use," Our World in Data, December 2019, https://ourworldindata.org/drug-use.

30 "deaths of despair": Anne Case and Angus Deaton, *Deaths of Despair and the Future of Capitalism* (Princeton, NJ: Princeton University Press, 2020), https://doi.org/10.2307/j.ctvpr7rb2.

30 world's natural resources are rapidly diminishing: "Capital Pains," *Economist*, July 18, 2020. For original sources see https://www .unenvironment.org/resources/report/inclusive-wealth-report-2018, and https://www.sciencedirect.com/science/article/pii /S0306261919305215.

35 "Religious man was born": Philip Rieff, *The Triumph of the Therapeutic: Uses of Faith after Freud* (New York: Harper and Row, 1966).

35 New Age "God Within" theology: Ross Douthat, *Bad Religion: How We Became a Nation of Heretics* (New York: Free Press, 2013).

38 pain was healthy: Maricia L. Meldrum, "A Capsule History of Pain Management," *JAMA* 290, no. 18 (2003): 2470–75, https://doi.org /10.1001/jama.290.18.2470.

38 opioids during surgery: Victoria K. Shanmugam, Kara S. Couch, Sean McNish, and Richard L. Amdur, "Relationship between Opioid Treatment and Rate of Healing in Chronic Wounds," *Wound Repair and Regeneration* 25, no. 1 (2017): 120–30, https://doi.org/10.1111/wrr.12496.

38 "instruments which nature makes use of": Thomas Sydenham, "A Treatise of the Gout and Dropsy," in *The Works of Thomas Sydenham, M.D., on Acute and Chronic Diseases* (London, 1783), 254, https://books.google.com/books?id=iSxsAAAAMAAJ&printsec=frontcover&source=gbs_ge_summary_r&cad=0#v=onepage&q&f=false.

38 massive prescribing of feel-good pills: Substance Abuse and Mental Health Services Administration, U.S. Department of Health and Human Services, "Behavioral Health, United States, 2012," HHS Publication No. (SMA) 13-4797, 2013, http://www.samhsa.gov/data/sites/default/files/2012-BHUS.pdf.

38 one in twenty American children: Bruce S. Jonas, Qiuping Gu, and Juan R. Albertorio-Diaz, "Psychotropic Medication Use among Adolescents: United States, 2005–2010," *NCHS Data Brief*, no. 135 (December 2013): 1–8.

38 use of antidepressants like Paxil, Prozac, and Celexa is rising: OECD, "OECD Health Statistics," July 2020, http://www.oecd.org/els/health-systems/health-data.htm. Laura A. Pratt, Debra J. Brody, Quiping Gu, "Antidepressant Use in Persons Aged 12 and Over: United States, 2005-2008," *NCHS Data Brief No. 76,* October 2011, https://www.cdc.gov/nchs/products/databriefs/db76.htm.

39 Prescriptions of stimulants (Adderall, Ritalin): Brian J. Piper, Christy L. Ogden, Olapeju M. Simoyan, Daniel Y. Chung, James F. Caggiano, Stephanie D. Nichols, and Kenneth L. McCall, "Trends in Use of Prescription Stimulants in the United States and Territories, 2006 to 2016," *PLOS ONE* 13, no. 11 (2018), https://doi.org/10.1371/journal.pone.0206100.

39 benzodiazepines (Xanax, Klonopin, Valium), also addictive, are on the rise: Marcus A. Bachhuber, Sean Hennessy, Chinazo O. Cunningham, and Joanna L. Starrels, "Increasing Benzodiazepine Prescriptions and Overdose Mortality in the United States, 1996–2013," *American Journal of Public Health* 106, no. 4 (2016): 686–88, https://doi.org/10.2105/AJPH.2016.303061.

40 **"infinite appetite for distractions":** Aldous Huxley, *Brave New World Revisited* (New York: HarperCollins, 2004).

40 **"Americans no longer talk to each other, they entertain each other":** Neil Postman, *Amusing Ourselves to Death: Public Discourse in the Age of Show Business* (New York: Penguin Books, 1986).

44 **World Happiness Report:** John F. Helliwell, Haifang Huang, and Shun Wang, "Chapter 2—Changing World Happiness," *World Happiness Report 2019*, March 20, 2019, 10–46.

45 **richer countries had higher rates of anxiety:** Ayelet Meron Ruscio, Lauren S. Hallion, Carmen C. W. Lim, Sergio Aguilar-Gaxiola, Ali Al-Hamzawi, Jordi Alonso, Laura Helena Andrade, et al., "Cross-Sectional Comparison of the Epidemiology of *DSM-5* Generalized Anxiety Disorder across the Globe," *JAMA Psychiatry* 74, no. 5 (2017): 465–75, https://doi.org/10.1001/jamapsychiatry.2017.0056.

45 **depression worldwide increased 50 percent:** Qingqing Liu, Hairong He, Jin Yang, Xiaojie Feng, Fanfan Zhao, and Jun Lyu, "Changes in the Global Burden of Depression from 1990 to 2017: Findings from the Global Burden of Disease Study," *Journal of Psychiatric Research* 126 (June 2019): 134–40, https://doi.org/10.1016/j.jpsychires.2019.08.002.

45 **Physical pain too is increasing:** David G. Blanchflower and Andrew J. Oswald, "Unhappiness and Pain in Modern America: A Review Essay, and Further Evidence, on Carol Graham's Happiness for All?" IZA Institute of Labor Economics discussion paper, November 2017.

46 **a time of unprecedented wealth:** Robert William Fogel, *The Fourth Great Awakening and the Future of Egalitarianism* (Chicago: University of Chicago Press, 2000).

48 **Kathleen Montagu, based outside of London:** Kathleen A. Montagu, "Catechol Compounds in Rat Tissues and in Brains of Different Animals," *Nature* 180 (1957): 244–45, https://doi.org/10.1038/180244a0.

49 *Wanting* **more than** *liking*: Bryon Adinoff, "Neurobiologic Processes in Drug Reward and Addiction," *Harvard Review of Psychiatry* 12, no. 6 (2004): 305–20, https://doi.org/10.1080/10673220490910844.

49 mice unable to make dopamine: Qun Yong Zhou and Richard D. Palmiter, "Dopamine-Deficient Mice Are Severely Hypoactive, Adipsic, and Aphagic," *Cell* 83, no. 7 (1995): 1197–1209, https://doi.org/10.1016/0092-8674(95)90145-0.

50 For a rat in a box, chocolate: Valentina Bassareo and Gaetano Di Chiara, "Modulation of Feeding-Induced Activation of Mesolimbic Dopamine Transmission by Appetitive Stimuli and Its Relation to Motivational State," *European Journal of Neuroscience* 11, no. 12 (1999): 4389–97, https://doi.org/10.1046/j.1460-9568.1999.00843.x.

50 sex by 100 percent: Dennis F. Fiorino, Ariane Coury, and Anthony G. Phillips, "Dynamic Changes in Nucleus Accumbens Dopamine Efflux during the Coolidge Effect in Male Rats," *Journal of Neuroscience* 17, no. 12 (1997): 4849–55, https://doi.org/10.1523/jneurosci.17-12-04849.1997.

50 nicotine by 150 percent: Gaetano Di Chiara and Assunta Imperato, "Drugs Abused by Humans Preferentially Increase Synaptic Dopamine Concentrations in the Mesolimbic System of Freely Moving Rats," *Proceedings of the National Academy of Sciences of the United States of America* 85, no. 14 (1988): 5274–78, https://doi.org/10.1073/pnas.85.14.5274.

50 overlapping brain regions: Siri Leknes and Irene Tracey, "A Common Neurobiology for Pain and Pleasure," *Nature Reviews Neuroscience* 9, no. 4 (2008): 314–20, https://doi.org/10.1038/nrn2333.

52 "departures from hedonic or affective neutrality": Richard L. Solomon and John D. Corbit, "An Opponent-Process Theory of Motivation," *American Economic Review* 68, no. 6 (1978): 12–24.

55 *opioid-induced hyperalgesia*: Yinghui Low, Collin F. Clarke, and Billy K. Huh, "Opioid-Induced Hyperalgesia: A Review of Epidemiology, Mechanisms and Management," *Singapore Medical Journal* 53, no. 5 (2012): 357–60.

55 when these patients tapered off opioids: Joseph W. Frank, Travis I. Lovejoy, William C. Becker, Benjamin J. Morasco, Christopher J. Koenig, Lilian Hoffecker, Hannah R. Dischinger, et al., "Patient Outcomes in Dose Reduction or Discontinuation of Long-Term Opioid Therapy: A Systematic Review," *Annals of Internal Medicine* 167, no. 3 (2017): 181–91, https://doi.org/10.7326/M17-0598.

56 "decreased sensitivity of reward circuits": Nora D. Volkow, Joanna S. Fowler, and Gene-Jack Wang, "Role of Dopamine in Drug Reinforcement and Addiction in Humans: Results from Imaging Studies," *Behavioural Pharmacology* 13, no. 5 (2002): 355–66, https://doi.org/10.1097/00008877-200209000-00008.

57 "dysphoria driven relapse": George F. Koob, "Hedonic Homeostatic Dysregulation as a Driver of Drug-Seeking Behavior," *Drug Discovery Today: Disease Models* 5, no. 4 (2008): 207–15, https://doi.org/10.1016/j.ddmod.2009.04.002.

62 gambling addiction: Jakob Linnet, Ericka Peterson, Doris J. Doudet, Albert Gjedde, and Arne Møller, "Dopamine Release in Ventral Striatum of Pathological Gamblers Losing Money," *Acta Psychiatrica Scandinavica* 122, no. 4 (2010): 326–33, https://doi.org/10.1111/j.1600-0447.2010.01591.x.

62 *experience-dependent plasticity*: Terry E. Robinson and Bryan Kolb, "Structural Plasticity Associated with Exposure to Drugs of Abuse," *Neuropharmacology* 47, Suppl. 1 (2004): 33–46, https://doi.org/10.1016/j.neuropharm.2004.06.025.

64 rat's ability to learn: Brian Kolb, Grazyna Gorny, Yilin Li, Anne-Noël Samaha, and Terry E. Robinson, "Amphetamine or Cocaine Limits the Ability of Later Experience to Promote Structural Plasticity in the Neocortex and Nucleus Accumbens," *Proceedings of the National Academy of Sciences of the United States of America* 100, no. 18 (2003): 10523–28, https://doi.org/10.1073/pnas.1834271100.

64 new synaptic pathways to create healthy behaviors: Sandra Chanraud, Anne-Lise Pitel, Eva M. Muller-Oehring, Adolf Pfefferbaum, and Edith V. Sullivan, "Remapping the Brain to Compensate for Impairment in Recovering Alcoholics," *Cerebral Cortex* 23 (2013): 97–104, doi:10.1093/cercor/bhr38; Changhai Cui, Antonio Noronha, Kenneth R. Warren, George F. Koob, Rajita Sinha, Mahesh Thakkar, John Matochik, et al., "Brain Pathways to Recovery from Alcohol Dependence," *Alcohol* 49, no. 5 (2015): 435–52. https://doi.org/10.1016/j.alcohol.2015.04.006.

64 optogenetics: Vincent Pascoli, Marc Turiault, and Christian Lüscher, "Reversal of Cocaine-Evoked Synaptic Potentiation Resets

Drug-Induced Adaptive Behaviour," *Nature* 481 (2012): 71–75, https://doi.org/10.1038/nature10709.

66 "a ticket to the safety": Henry Beecher, "Pain in Men Wounded in Battle," *Anesthesia & Analgesia*, 1947, https://doi.org/10.1213/00000539-194701000-00005.

66 footfirst on a fifteen-centimeter nail: J. P. Fisher, D. T. Hassan, and N. O'Connor, "Case Report on Pain," *British Medical Journal* 310, no. 6971 (1995): 70, https://www.ncbi.nlm.nih.gov/pmc/articles/PMC2548478/pdf/bmj00574-0074.pdf.

67 "We are cacti in the rain forest": Dr. Tom Finucane is a professor of medicine at Johns Hopkins in Baltimore, whose work I came across when I was lecturing there on a visiting professorship. It was during a dinner with some of his students that I first heard this phrase, and I knew I had to find a way to include it in this book.

78 dopamine transmission is still below normal: Nora D. Volkow, Joanna S. Fowler, Gene-Jack Wang, and James M. Swanson, "Dopamine in Drug Abuse and Addiction: Results from Imaging Studies and Treatment Implications," *Molecular Psychiatry* 9, no. 6 (June 2004): 557–69, https://doi.org/10.1038/sj.mp.4001507.

78 After one month of not drinking: Sandra A. Brown and Marc A. Schuckit, "Changes in Depression among Abstinent Alcoholics," *Journal on Studies of Alcohol* 49, no. 5 (1988): 412–17, https://pubmed.ncbi.nlm.nih.gov/3216643/.

79 standard treatments for depression: Kenneth B. Wells, Roland Sturm, Cathy D. Sherbourne, and Lisa S. Meredith, *Caring for Depression* (Cambridge, MA: Harvard University Press, 1996).

86 using their drug of choice in a controlled way: Mark B. Sobell and Linda C. Sobell, "Controlled Drinking after 25 Years: How Important Was the Great Debate?," *Addiction* 90, no. 9 (1995): 1149–53. Linda C. Sobell, John A. Cunningham, and Mark B. Sobell, "Recovery from Alcohol Problems with and without Treatment: Prevalence in Two Population Surveys," *American Journal of Public Health* 86, no. 7 (1996): 966–72.

87 *abstinence violation effect*: Roelof Eikelboom and Randelle Hewitt, "Intermittent Access to a Sucrose Solution for Rats Causes

Long-Term Increases in Consumption," *Physiology and Behavior* 165 (2016): 77–85, https://doi.org/10.1016/j.physbeh.2016.07.002.

87 binge on alcohol as soon as they have access: Valentina Vengeliene, Ainhoa Bilbao, and Rainer Spanagel, "The Alcohol Deprivation Effect Model for Studying Relapse Behavior: A Comparison between Rats and Mice," *Alcohol* 48, no. 3 (2014): 313–20, https://doi.org/10.1016/j.alcohol.2014.03.002.

91 *Self-binding* is the term to describe: I first came across the term *self-binding* in this article by Sally Satel and Scott O. Lilienfeld. Sally Satel and Scott O. Lilienfeld, "Addiction and the Brain-Disease Fallacy," *Frontiers in Psychiatry* 4 (March 2014): 1–11, https://doi.org /10.3389/fpsyt.2013.00141. I've been a fan of Satel's work for some time, and here she was using *self-binding* to emphasize "the vast role of personal agency in perpetuating the cycle of use and relapse." But I disagree with the basic premise of this article, which argues that our ability to self-bind refutes the disease model of addiction. For me, our need to self-bind speaks to the powerful pull of addiction and the brain changes that go along with it, consistent with the disease model. Economist Thomas Schelling also addresses the concept of self-binding, but calls it "self-management" and "self-command": "Self-Command in Practice, in Policy, and in a Theory of Rational Choice," *American Economic Review* 74, no. 2 (1984): 1–11, https://econpapers .repec.org/article/aeaaecrev/v_3a74_3ay_3a1984_3ai_3a2_3ap_3a1-11 .htm. https://doi.org/10.2307/1816322.

96 naltrexone half an hour before: J. D. Sinclair, "Evidence about the Use of Naltrexone and for Different Ways of Using It in the Treatment of Alcoholism," *Alcohol and Alcoholism* 36, no. 1 (2001): 2–10, https://doi.org/10.1093/alcalc/36.1.2.

96 addiction treatment hospital in Beijing: Anna Lembke and Niushen Zhang, "A Qualitative Study of Treatment-Seeking Heroin Users in Contemporary China," *Addiction Science & Clinical Practice* 10, no. 23 (2015), https://doi.org/10.1186/s13722-015-0044-3.

98 disulfiram-like reaction to alcohol: Jeffrey S. Chang, Jenn Ren Hsiao, and Che Hong Chen, "ALDH2 Polymorphism and Alcohol-Related Cancers in Asians: A Public Health Perspective," *Journal of Biomedical Science* 24, no. 19 (2017): 1–10, https://doi.org /10.1186/s12929-017-0327-y.

99 gastric bypass surgery . . . a new problem with alcohol:
Magdalena Plecka Östlund, Olof Backman, Richard Marsk, Dag
Stockeld, Jesper Lagergren, Finn Rasmussen, and Erik Näslund,
"Increased Admission for Alcohol Dependence after Gastric Bypass
Surgery Compared with Restrictive Bariatric Surgery," *JAMA Surgery*
148, no. 4 (2013): 374–77, https://doi.org/10.1001/jamasurg.2013.700.

101 extended access . . . methamphetamine: Jason L. Rogers, Silvia
De Santis, and Ronald E. See, "Extended Methamphetamine Self-
Administration Enhances Reinstatement of Drug Seeking and
Impairs Novel Object Recognition in Rats," *Psychopharmacology* 199,
no. 4 (2008): 615–24, https://doi.org/10.1007/s00213-008-1187-7.

101 extended access . . . nicotine: Laura E. O'Dell, Scott A. Chen,
Ron T. Smith, Sheila E. Specio, Robert L. Balster, Neil E. Paterson,
Athina Markou, et al., "Extended Access to Nicotine Self-
Administration Leads to Dependence: Circadian Measures,
Withdrawal Measures, and Extinction Behavior in Rats," *Journal of
Pharmacology and Experimental Therapeutics* 320, no. 1 (2007):
180–93, https://doi.org/10.1124/jpet.106.105270.

101 extended access . . . heroin: Scott A. Chen, Laura E. O'Dell,
Michael E. Hoefer, Thomas N. Greenwell, Eric P. Zorrilla, and George
F. Koob, "Unlimited Access to Heroin Self-Administration:
Independent Motivational Markers of Opiate Dependence,"
Neuropsychopharmacology 31, no. 12 (2006): 2692–707, https://doi.org/
10.1038/sj.npp.1301008.

101 extended access . . . alcohol: Marcia Spoelder, Peter Hesseling,
Annemarie M. Baars, José G. Lozeman-van't Klooster, Marthe D.
Rotte, Louk J. M. J. Vanderschuren, and Heidi M. B. Lesscher,
"Individual Variation in Alcohol Intake Predicts Reinforcement,
Motivation, and Compulsive Alcohol Use in Rats," *Alcoholism: Clinical
and Experimental Research* 39, no. 12 (2015): 2427–37, https://doi.org
/10.1111/acer.12891.

101 steady amounts of cocaine: Serge H. Ahmed and George F. Koob,
"Transition from Moderate to Excessive Drug Intake: Change in
Hedonic Set Point," *Science* 282, no. 5387 (1998): 298–300, https://doi
.org/10.1126/science.282.5387.298.

103 a winning lottery ticket: Anne L. Bretteville-Jensen, "Addiction and Discounting," *Journal of Health Economics* 18, no. 4 (1999): 393–407, https://doi.org/10.1016/S0167-6296(98)00057-5.

103 Cigarette smokers . . . discount future rewards: Warren K. Bickel, Benjamin P. Kowal, and Kirstin M. Gatchalian, "Understanding Addiction as a Pathology of Temporal Horizon," *Behavior Analyst Today* 7, no. 1 (2006): 32–47, https://doi.org/10.1037/h0100148.

104 "temporal horizons" shrink: Nancy M. Petry, Warren K. Bickel, and Martha Arnett, "Shortened Time Horizons and Insensitivity to Future Consequences in Heroin Addicts," *Addiction* 93, no. 5 (1998): 729–38, https://doi.org/10.1046/j.1360-0443.1998.9357298.x.

104 immediate versus delayed rewards: Samuel M. McClure, David I. Laibson, George Loewenstein, and Jonathan D. Cohen, "Separate Neural Systems Value Immediate and Delayed Monetary Rewards," *Science* 306, no. 5695 (2004): 503–7, https://doi.org/10.1126/science.1100907.

105 Young Brazilians living in favelas: Dandara Ramos, Tânia Victor, Maria L. Seidl-de-Moura, and Martin Daly, "Future Discounting by Slum-Dwelling Youth versus University Students in Rio de Janeiro," *Journal of Research on Adolescence* 23, no. 1 (2013): 95–102, https://doi.org/10.1111/j.1532-7795.2012.00796.x.

106 the amount of leisure time in: Robert William Fogel, *The Fourth Great Awakening and the Future of Egalitarianism* (Chicago: University of Chicago Press, 2000). These data on leisure and work in the United States come from Fogel's book, an awe-inspiring analysis of the economic, social, and spiritual transformation in the United States in the last four hundred years.

106 high-income countries are similar: OECD, "Special Focus: Measuring Leisure in OECD Countries," in *Society at a Glance 2009: OECD Social Indicators* (Paris: OECD Publishing, 2009), https://doi.org/10.1787/soc_glance-2008-en.

106 differs by education and socioeconomic status: David R. Francis, "Why High Earners Work Longer Hours," National Bureau of

Economic Research digest, September 2020, http://www.nber.org /digest/jul06/w11895.html.

107 **"shifted their leisure to video gaming"**: Mark Aguiar, Mark Bils, Kerwin K. Charles, and Erik Hurst, "Leisure Luxuries and the Labor Supply of Young Men," National Bureau of Economic Research working paper, June 2017, https://doi.org/10.3386/w23552.

107 **"bored or frustrated problem-solvers"**: Eric J. Iannelli, "Species of Madness," *Times Literary Supplement*, September 22, 2017.

112 **"women to cast down their glances"**: "Qur'an: Verse 24:31," accessed July 2, 2020, http://corpus.quran.com/translation.jsp? chapter=24&verse=31.

112 **"short shorts and short skirts"**: The Church of Jesus Christ of Latter-day Saints, "Dress and Appearance," accessed July 2, 2020, https://www.churchofjesuschrist.org/study/manual/for-the-strength -of-youth/dress-and-appearance?lang=eng.

113 **3,000 new gluten-free snack products**: M. Shahbandeh, "Gluten-Free Food Market Value in the United States from 2014 to 2025," Statista, November 20, 2019, accessed July 2, 2020, https://www .statista.com/statistics/884086/us-gluten-free-food-market-value/.

115 **famous Stanford marshmallow experiment**: Yuichi Shoda, Walter Mischel, and Philip K. Peake, "Predicting Adolescent Cognitive and Self-Regulatory Competencies from Preschool Delay of Gratification: Identifying Diagnostic Conditions," *Developmental Psychology* 26, no. 6 (1990): 978–86, https://doi.org/10.1037/0012-1649.26.6.978.

116 **"cover their eyes with their hands"**: Roy F. Baumeister, "Where Has Your Willpower Gone?," *New Scientist* 213, no. 2849 (2012): 30–31, https://doi.org/10.1016/s0262-4079(12)60232-2.

118 **"reverence for the moral man"**: Immanuel Kant, "Groundwork of the Metaphysic of Morals (1785)," *Cambridge Texts in the History of Philosophy* (Cambridge: Cambridge University Press, 1998).

119 **buprenorphine decreases illicit opioid use:** John Strang, Thomas Babor, Jonathan Caulkins, Benedikt Fischer, David Foxcroft, and Keith Humphreys, "Drug Policy and the Public Good: Evidence for Effective Interventions," *Lancet* 379 (2012): 71-83.

124 **doctors in Arkansas wrote 116 opioid prescriptions**: Centers for Disease Control and Prevention, "U.S. Opioid Prescribing Rate

Maps," accessed July 2, 2020, https://www.cdc.gov/drugoverdose/maps/rxrate-maps.html.

129 evidence for psychotropic medications more generally is not robust: Robert Whitaker, *Anatomy of an Epidemic: Magic Bullets, Psychiatric Drugs, and the Astonishing Rise of Mental Illness in America* (New York: Crown, 2010).

129 Despite substantial increases in funding . . . for psychiatric medications: Anthony F. Jorm, Scott B. Patten, Traolach S. Brugha, and Ramin Mojtabai, "Has Increased Provision of Treatment Reduced the Prevalence of Common Mental Disorders? Review of the Evidence from Four Countries," *World Psychiatry* 16, no. 1 (2017): 90–99, https://doi.org/10.1002/wps.20388.

130 process called opioid-induced hyperalgesia: Larry F. Chu, David J. Clark, and Martin S. Angst, "Opioid Tolerance and Hyperalgesia in Chronic Pain Patients after One Month of Oral Morphine Therapy: A Preliminary Prospective Study," *Journal of Pain* 7, no. 1 (2006): 43–48, https://doi.org/10.1016/j.jpain.2005.08.001.

130 "ADHD drug treatment is associated with deterioration": Gretchen LeFever Watson, Andrea Powell Arcona, and David O. Antonuccio, "The ADHD Drug Abuse Crisis on American College Campuses," *Ethical Human Psychology and Psychiatry* 17, no. 1 (2015), https://doi.org/10.1891/1559-4343.17.1.5.

130 *tardive dysphoria*: Rif S. El-Mallakh, Yonglin Gao, and R. Jeannie Roberts, "Tardive Dysphoria: The Role of Long Term Antidepressant Use in-Inducing Chronic Depression," *Medical Hypotheses* 76, no. 6 (2011): 769–73, https://doi.org/10.1016/j.mehy.2011.01.020.

131 antidepressants make people "better than well": Peter D. Kramer, *Listening to Prozac* (New York: Viking Press, 1993).

133 7.5 percent of American children: Lajeana D. Howie, Patricia N. Pastor, and Susan L. Lukacs, "Use of Medication Prescribed for Emotional or Behavioral Difficulties among Children Aged 6–17 Years in the United States, 2011–2012," *Health Care in the United States: Developments and Considerations* 5, no. 148 (2015): 25–35.

133 as many as ten thousand toddlers: Alan Schwarz, "Thousands of Toddlers Are Medicated for A.D.H.D., Report Finds, Raising Worries," *New York Times*, May 16, 2014.

133 "reaction to adverse and inhumane treatment": Edmund C. Levin, "The Challenges of Treating Developmental Trauma Disorder in a Residential Agency for Youth," *Journal of the American Academy of Psychoanalysis and Dynamic Psychiatry* 37, no. 3 (2009): 519–38, https://doi.org/10.1521/jaap.2009.37.3.519.

134 "neighborhood deprivation is associated": Casey Crump, Kristina Sundquist, Jan Sundquist, and Marilyn A. Winkleby, "Neighborhood Deprivation and Psychiatric Medication Prescription: A Swedish National Multilevel Study," *Annals of Epidemiology* 21, no. 4 (2011): 231–37, https://doi.org/10.1016/j.annepidem.2011.01.005.

134 "counties with worse economic prospects": Robin Ghertner and Lincoln Groves, "The Opioid Crisis and Economic Opportunity: Geographic and Economic Trends," ASPE Research Brief from the U.S. Department of Health and Human Services, 2018, https://aspe.hhs.gov/system/files/pdf/259261/ASPEEconomicOpportunityOpioidCrisis.pdf.

134 Medicaid patients die from opioids: Mark J. Sharp and Thomas A. Melnik, "Poisoning Deaths Involving Opioid Analgesics—New York State, 2003–2012," *Morbidity and Mortality Weekly Report* 64, no. 14 (2015): 377–80; P. Coolen, S. Best, A. Lima, J. Sabel, and L. J. Paulozzi, "Overdose Deaths Involving Prescription Opioids among Medicaid Enrollees—Washington, 2004–2007," *Morbidity and Mortality Weekly* Report 58, no. 42 (2009): 1171–75.

134 "medication alone, rather than being liberatory": Alexandrea E. Hatcher, Sonia Mendoza, and Helena Hansen, "At the Expense of a Life: Race, Class, and the Meaning of Buprenorphine in Pharmaceuticalized 'Care,'" *Substance Use and Misuse* 53, no. 2 (2018): 301–10, https://doi.org/10.1080/10826084.2017.1385633.

142 ten men volunteered to submerge themselves: Petr Šrámek, Marie Šimečková, Ladislav Janský, Jarmila Šavlíková, and Stanislav Vybíral, "Human Physiological Responses to Immersion into Water of Different Temperatures," *European Journal of Applied Physiology* 81 (2000): 436–42, https://doi.org/10.1007/s004210050065.

143 brains of hibernating ground squirrels: Christina G. von der Ohe, Corinna Darian-Smith, Craig C. Garner, and H. Craig Heller, "Ubiquitous and Temperature-Dependent Neural Plasticity in Hibernators," *Journal of Neuroscience* 26, no. 41 (2006): 10590–98, https://doi.org/10.1523/JNEUROSCI.2874-06.2006.

145 a series of experiments on dogs: Russell M. Church, Vincent LoLordo, J. Bruce Overmier, Richard L. Solomon, and Lucille H. Turner, "Cardiac Responses to Shock in Curarized Dogs: Effects of Shock Intensity and Duration, Warning Signal, and Prior Experience with Shock," *Journal of Comparative and Physiological Psychology* 62, no. 1 (1966): 1–7, https://doi.org/10.1037/h0023476; Aaron H. Katcher, Richard L. Solomon, Lucille H. Turner, Vincent LoLordo, J. Bruce Overmier, and Robert A. Rescorla, "Heart Rate and Blood Pressure Responses to Signaled and Unsignaled Shocks: Effects of Cardiac Sympathectomy," *Journal of Comparative and Physiological Psychology* 68, no. 2 (1969): 163–74; Richard L. Solomon and John D. Corbit, "An Opponent-Process Theory of Motivation," *American Economic Review* 68, no. 6 (1978): 12–24.

147 "this thing that men call pleasure!": R. S. Bluck, *Plato's* Phaedo: *A Translation of Plato's* Phaedo (London: Routledge, 2014), https://www.google.com/books/edition/Plato_s_Phaedo/7FzXAwAAQBAJ?hl=en&gbpv=1&dq=%22how+strange+would+appear+to+be+this+thing+that+men+call+pleasure%22&pg=PA41&printsec=frontcover.

147 "son was struck by lightning": Helen B. Taussig, "'Death' from Lightning and the Possibility of Living Again," *American Scientist* 57, no. 3 (1969): 306–16.

148 "environmental or self-imposed challenges": Edward J. Calabrese and Mark P. Mattson, "How Does Hormesis Impact Biology, Toxicology, and Medicine?," *npj Aging and Mechanisms of Disease* 3, no. 13 (2017), https://doi.org/10.1038/s41514-017-0013-z.

148 Worms exposed to temperatures: James R. Cypser, Pat Tedesco, and Thomas E. Johnson, "Hormesis and Aging in *Caenorhabditis Elegans*," *Experimental Gerontology* 41, no. 10 (2006): 935–39, https://doi.org/10.1016/j.exger.2006.09.004.

149 Fruit flies that were spun in a centrifuge: Nadège Minois, "The Hormetic Effects of Hypergravity on Longevity and Aging," *Dose-Response* 4, no. 2 (2006), https://doi.org/10.2203/dose-response.05-008.minois. When I read this study, I imagined spending two to four weeks in a Gravitron at my local amusement park, the large, upright barrel that rotates at 33 revolutions per minute, creating a centrifugal effect equivalent to almost 3 g's before the floor drops out. Given that the average life span of the fruit fly is fifty days, this amounts to more than fifty human years in the Gravitron. Those poor flies!

149 "stimulated anticancer immunity": Shizuyo Sutou, "Low-Dose Radiation from A-Bombs Elongated Lifespan and Reduced Cancer Mortality Relative to Un-Irradiated Individuals," *Genes and Environment* 40, no. 26 (2018), https://doi.org/10.1186/s41021-018-0114-3.

149 these findings are controversial: John B. Cologne and Dale L. Preston, "Longevity of Atomic-Bomb Survivors," *Lancet* 356, no. 9226 (July 22, 2000): 303–7, https://doi.org/10.1016/S0140-6736(00)02506-X.

149 calorie restriction extended lifespan: Mark P. Mattson and Ruiqian Wan, "Beneficial Effects of Intermittent Fasting and Caloric Restriction on the Cardiovascular and Cerebrovascular Systems," *Journal of Nutritional Biochemistry* 16, no. 3 (2005): 129–37, https://doi.org/10.1016/j.jnutbio.2004.12.007.

150 "starving myself two days a week": Aly Weisman and Kristen Griffin, "Jimmy Kimmel Lost a Ton of Weight on This Radical Diet," *Business Insider,* January 9, 2016.

150 Exercise increases many of the neurotransmitters: Anna Lembke and Amer Raheemullah, "Addiction and Exercise," in *Lifestyle Psychiatry: Using Exercise, Diet and Mindfulness to Manage Psychiatric Disorders*, ed. Doug Noordsy (Washington, DC: American Psychiatric Publishing, 2019).

151 Dopamine's ancient role in physical movement: Daniel T. Omura, Damon A. Clark, Aravinthan D. T. Samuel, and H. Robert Horvitz, "Dopamine Signaling Is Essential for Precise Rates of Locomotion by *C. Elegans*," *PLOS ONE* 7, no. 6 (2012), https://doi.org/10.1371/journal.pone.0038649.

151 half their waking hours sitting: Shu W. Ng and Barry M. Popkin, "Time Use and Physical Activity: A Shift Away from Movement across the Globe," *Obesity Reviews* 13, no. 8 (August 2012): 659–80, https://doi.org/10.1111/j.1467-789X.2011.00982.x.

151 traverse tens of kilometers daily: Mark P. Mattson, "Energy Intake and Exercise as Determinants of Brain Health and Vulnerability to Injury and Disease," *Cell Metabolism* 16, no. 6 (2012): 706–22, https://doi.org/10.1016/j.cmet.2012.08.012.

152 Exercise has a more . . . than any pill I can prescribe: B. K. Pedersen and B. Saltin, "Exercise as Medicine—Evidence for Prescribing Exercise as Therapy in 26 Different Chronic Diseases,"

Scandinavian Journal of Medicine and Science in Sports 25, no. S3 (2015): 1–72.

152 "tyranny of convenience": Tim Wu, "The Tyranny of Convenience," *New York Times,* February 6, 2018.

153 "Of two pains occurring together": Hippocrates, *Aphorisms,* accessed July 8, 2020, http://classics.mit.edu/Hippocrates/aphorisms .1.i.html.

153 application of a second painful stimulus: Christian Sprenger, Ulrike Bingel, and Christian Büchel, "Treating Pain with Pain: Supraspinal Mechanisms of Endogenous Analgesia Elicited by Heterotopic Noxious Conditioning Stimulation," *Pain* 152, no. 2 (2011): 428–39, https://doi.org/10.1016/j.pain.2010.11.018.

154 "The needling, which can injure": Liu Xiang, "Inhibiting Pain with Pain—A Basic Neuromechanism of Acupuncture Analgesia," *Chinese Science Bulletin* 46, no. 17 (2001): 1485–94, https://doi.org/10 .1007/BF03187038.

155 "greater reduction in their pain scores": Jarred Younger, Noorulain Noor, Rebecca McCue, and Sean Mackey, "Low-Dose Naltrexone for the Treatment of Fibromyalgia: Findings of a Small, Randomized, Double-Blind, Placebo-Controlled, Counterbalanced, Crossover Trial Assessing Daily Pain Levels," *Arthritis and Rheumatism* 65, no. 2 (2013): 529–38, https://doi.org/10.1002/art.37734.

155 "incomprehensible gibberish made up of odd neologisms": Ugo Cerletti, "Old and New Information about Electroshock," *American Journal of Psychiatry* 107, no. 2 (1950): 87–94, https://doi.org /10.1176/ajp.107.2.87.

156 "ECT brings about various": Amit Singh and Sujita Kumar Kar, "How Electroconvulsive Therapy Works?: Understanding the Neurobiological Mechanisms," *Clinical Psychopharmacology and Neuroscience* 15, no. 3 (2017): 210–21, https://doi.org/10.9758 /cpn.2017.15.3.210.

159 "I've done so much soloing": Mark Synnott, *The Impossible Climb: Alex Honnold, El Capitan, and the Climbing Life* (New York: Dutton, 2018).

163 rats run until they die: Chris M. Sherwin, "Voluntary Wheel Running: A Review and Novel Interpretation," *Animal Behaviour* 56, no. 1 (1998): 11–27, https://doi.org/10.1006/anbe.1998.0836.

164 "feral mice ran in the wheels year-round": Johanna H. Meijer and Yuri Robbers, "Wheel Running in the Wild," *Proceedings of the Royal Society B: Biological Sciences*, July 7, 2014, https://doi.org/10.1098/rspb.2014.0210.

165 stress alone can increase the release of dopamine: Daniel Saal, Yan Dong, Antonello Bonci, and Robert C. Malenka, "Drugs of Abuse and Stress Trigger a Common Synaptic Adaptation in Dopamine Neurons," *Neuron* 37, no. 4 (2003): 577–82, https://doi.org/10.1016/S0896-6273(03)00021-7.

165 "skydiving has similarities with addictive behaviors": Ingmar H. A. Franken, Corien Zijlstra, and Peter Muris, "Are Nonpharmacological Induced Rewards Related to Anhedonia? A Study among Skydivers," *Progress in Neuro-Psychopharmacology and Biological Psychiatry* 30, no. 2 (2006): 297–300, https://doi.org/10.1016/j.pnpbp.2005.10.011.

166 "ice cooler to cool my core down": Kate Knibbs, "All the Gear an Ultramarathoner Legend Brings with Him on the Trail," Gizmodo, October 29, 2015, https://gizmodo.com/all-the-gear-an-ultramarathon-legend-brings-with-him-on-1736088954.

166 "memorizing thousands of intricate hand and foot sequences": Mark Synnott, "How Alex Honnold Made the Ultimate Climb without a Rope," *National Geographic* online, accessed July 8, 2020, https://www.nationalgeographic.com/magazine/2019/02/alex-honnold-made-ultimate-climb-el-capitan-without-rope.

167 "Overtraining syndrome": Jeffrey B. Kreher and Jennifer B. Schwartz, "Overtraining Syndrome: A Practical Guide," *Sports Health* 4, no. 2 (2012), https://doi.org/10.1177/1941738111434406.

169 highly educated wage earners are working more: David R. Francis, "Why High Earners Work Longer Hours," National Bureau of Economic Research digest, accessed February 5, 2021, https://www.nber.org/digest/jul06/w11895.html.

172 average adult tells between 0.59 and 1.56 lies daily: Silvio José Lemos Vasconcellos, Matheus Rizzatti, Thamires Pereira Barbosa, Bruna Sangoi Schmitz, Vanessa Cristina Nascimento Coelho, and Andrea Machado, "Understanding Lies Based on Evolutionary Psychology: A Critical Review," *Trends in Psychology* 27, no. 1 (2019): 141–53, https://doi.org/10.9788/TP2019.1-11.

177 the neurobiological mechanisms of honesty: Michel André Maréchal, Alain Cohn, Giuseppe Ugazio, and Christian C. Ruff, "Increasing Honesty in Humans with Noninvasive Brain Stimulation," *Proceedings of the National Academy of Sciences of the United States of America* 114, no. 17 (2017): 4360–64, https://doi.org /10.1073/pnas.1614912114.

184 oxytocin leads to an increase in brain dopamine: Oxytocin also causes release of serotonin (5HT) in the major dopamine target—the nucleus accumbens—and it's the release of serotonin in the nucleus accumbens that is more important than the release of dopamine for promoting "prosocial" behaviors. The concurrent release of dopamine, however, is probably what makes prosocial behaviors potentially addictive. Lin W. Hung, Sophie Neuner, Jai S. Polepalli, Kevin T. Beier, Matthew Wright, Jessica J. Walsh, Eastman M. Lewis, et al., "Gating of Social Reward by Oxytocin in the Ventral Tegmental Area," *Science* 357, no. 6358 (2017): 1406–11, https://doi.org/10.1126/science.aan4994.

184 rat trapped inside a plastic bottle: Seven E. Tomek, Gabriela M. Stegmann, and M. Foster Olive, "Effects of Heroin on Rat Prosocial Behavior," *Addiction Biology* 24, no. 4 (2019): 676–84, https://doi.org /10.1111/adb.12633.

187 AA philosophy and teachings: *Twelve Steps and Twelve Traditions* (New York: Alcoholics Anonymous World Services).

191 concept of "the false self": Donald W. Winnicott, "Ego Distortion in Terms of True and False Self," in *The Maturational Process and the Facilitating Environment: Studies in the Theory of Emotional Development* (New York: International Universities Press, 1960), 140–57.

192 "a feeling of connection": Mark Epstein, *Going on Being: Life at the Crossroads of Buddhism and Psychotherapy* (Boston: Wisdom Publications, 2009).

194 children experienced a broken promise: Celeste Kidd, Holly Palmeri, and Richard N. Aslin, "Rational Snacking: Young Children's Decision-Making on the Marshmallow Task Is Moderated by Beliefs about Environmental Reliability," *Cognition* 126, no. 1 (2013): 109–14, https://doi.org/10.1016/j.cognition.2012.08.004.

195 "just been fired from your job": Warren K. Bickel, A. George Wilson, Chen Chen, Mikhail N. Koffarnus, and Christopher T. Franck, "Stuck in Time: Negative Income Shock Constricts the Temporal

Window of Valuation Spanning the Future and the Past," *PLOS ONE* 11, no. 9 (2016): 1–12, https://doi.org/10.1371/journal.pone.0163051.

214 actively involved in religious organizations: Mark J. Edlund, Katherine M. Harris, Harold G. Koenig, Xiaotong Han, Greer Sullivan, Rhonda Mattox, and Lingqi Tang, "Religiosity and Decreased Risk of Substance Use Disorders: Is the Effect Mediated by Social Support or Mental Health Status?," *Social Psychiatry and Psychiatric Epidemiology* 45 (2010): 827–36, https://doi.org/10.1007/s00127-009-0124-3.

219 "pleasure I derive from Sunday service": Laurence R. Iannaccone, "Sacrifice and Stigma: Reducing Free-Riding in Cults, Communes, and Other Collectives," *Journal of Political Economy* 100, no. 2 (1992): 271–91.

220 "stigmatizing behaviors that reduce participation": Laurence R. Iannaccone, "Why Strict Churches Are Strong," *American Journal of Sociology* 99, no. 5 (1994): 1180–1211, https://doi.org/10.2307/2781147.

Bibliography

Adinoff, Bryon. "Neurobiologic Processes in Drug Reward and Addiction." *Harvard Review of Psychiatry* 12, no. 6 (2004): 305–20. https://doi.org/10.1080/10673220490910844.

Aguiar, Mark, Mark Bils, Kerwin Kofi Charles, and Erik Hurst. "Leisure Luxuries and the Labor Supply of Young Men." National Bureau of Economic Research working paper, June 2017. https://doi.org/10.3386/w23552.

Ahmed, S. H., and G. F. Koob. "Transition from Moderate to Excessive Drug Intake: Change in Hedonic Set Point." *Science* 282, no. 5387 (1998): 298–300. https://doi.org/10.1126/science.282.5387.298.

ASPPH Task Force on Public Health Initiatives to Address the Opioid Crisis. *Bringing Science to Bear on Opioids: Report and Recommendations*, November 2019.

Bachhuber, Marcus A., Sean Hennessy, Chinazo O. Cunningham, and Joanna L. Starrels. "Increasing Benzodiazepine Prescriptions and Overdose Mortality in the United States, 1996–2013." *American Journal of Public Health* 106, no. 4 (2016): 686–88. https://doi.org/10.2105/AJPH.2016.303061.

Bassareo, Valentina, and Gaetano Di Chiara. "Modulation of Feeding-Induced Activation of Mesolimbic Dopamine Transmission by Appetitive Stimuli and Its Relation to Motivational State." *European Journal of Neuroscience* 11, no. 12 (1999): 4389–97. https://doi.org/10.1046/j.1460-9568.1999.00843.x.

Baumeister, Roy F. "Where Has Your Willpower Gone?" *New Scientist* 213, no. 2849 (2012): 30–31. https://doi.org/10.1016/s0262-4079(12)60232-2.

Beecher, Henry. "Pain in Men Wounded in Battle." *Anesthesia & Analgesia* 26, no. 1 (1947): 21. https://doi.org/10.1213/00000539 -194701000-00005.

Bickel, Warren K., A. George Wilson, Chen Chen, Mikhail N. Koffarnus, and Christopher T. Franck. "Stuck in Time: Negative Income Shock Constricts the Temporal Window of Valuation Spanning the Future and the Past." *PLOS ONE* 11, no. 9 (2016): 1–12. https://doi.org/10.1371/journal.pone.0163051.

Bickel, Warren K., Benjamin P. Kowal, and Kirstin M. Gatchalian. "Understanding Addiction as a Pathology of Temporal Horizon." *Behavior Analyst Today* 7, no. 1 (2006): 32–47. https://doi.org/10.1037 /h0100148.

Blanchflower, David G., and Andrew J. Oswald. "Unhappiness and Pain in Modern America: A Review Essay, and Further Evidence, on Carol Graham's Happiness for All?" IZA Institute of Labor Economics discussion paper, November 2017.

Bluck, R. S. *Plato's* Phaedo: *A Translation of Plato's* Phaedo. London: Routledge, 2014. https://www.google.com/books/edition/Plato_s _Phaedo/7FzXAwAAQBAJ?hl=en&gbpv=1&dq=%22how+strange +would+appear+to+be+this+thing+that+men+call+pleasure%22&pg =PA41&printsec=frontcover.

Bretteville-Jensen, A. L. "Addiction and Discounting." *Journal of Health Economics* 18, no. 4 (1999): 393–407. https://doi.org/10.1016 /S0167-6296(98)00057-5.

Brown, S. A., and M. A. Schuckit. "Changes in Depression among Abstinent Alcoholics." *Journal on Studies of Alcohol* 49, no. 5 (1988): 412–17. http://www.ncbi.nlm.nih.gov/entrez/query.fcgi?cmd =Retrieve&db=PubMed&dopt=Citation&list_uids=3216643.

Calabrese, Edward J., and Mark P. Mattson. "How Does Hormesis Impact Biology, Toxicology, and Medicine?" *npj Aging and Mechanisms of Disease* 3, no. 13 (2017). https://doi.org/10.1038/s41514-017-0013-z.

"Capital Pains." *Economist*, July 18, 2020.

Case, Anne, and Angus Deaton. *Deaths of Despair and the Future of Capitalism*. Princeton, NJ: Princeton University Press, 2020. https://doi.org/10.2307/j.ctvpr7rb2.

Centers for Disease Control and Prevention. "U.S. Opioid Prescribing Rate Maps." Accessed July 2, 2020. https://www.cdc.gov /drugoverdose/maps/rxrate-maps.html.

Cerletti, Ugo. "Old and New Information about Electroshock." *American Journal of Psychiatry* 107, no. 2 (1950): 87–94. https://doi .org/10.1176/ajp.107.2.87.

Chang, Jeffrey S., Jenn Ren Hsiao, and Che Hong Chen. "ALDH2 Polymorphism and Alcohol-Related Cancers in Asians: A Public Health Perspective." *Journal of Biomedical Science* 24, no. 1 (2017): 1–10. https://doi.org/10.1186/s12929-017-0327-y.

Chanraud, Sandra, Anne-Lise Pitel, Eva M. Muller-Oehring, Adolf Pfefferbaum, and Edith V. Sullivan. "Remapping the Brain to Compensate for Impairment in Recovering Alcoholics," *Cerebral Cortex* 23 (2013): 97–104. http://doi:10.1093/cercor/bhr38.

Chen, Scott A., Laura E. O'Dell, Michael E. Hoefer, Thomas N. Greenwell, Eric P. Zorrilla, and George F. Koob. "Unlimited Access to Heroin Self-Administration: Independent Motivational Markers of Opiate Dependence." *Neuropsychopharmacology* 31, no. 12 (2006): 2692–707. https://doi.org/10.1038/sj.npp.1301008.

Chiara, G. Di, and A. Imperato. "Drugs Abused by Humans Preferentially Increase Synaptic Dopamine Concentrations in the Mesolimbic System of Freely Moving Rats." *Proceedings of the National Academy of Sciences of the United States of America* 85, no. 14 (1988): 5274–78. https://doi.org/10.1073/pnas.85.14.5274.

Chu, Larry F., David J. Clark, and Martin S. Angst. "Opioid Tolerance and Hyperalgesia in Chronic Pain Patients after One Month of Oral Morphine Therapy: A Preliminary Prospective Study." *Journal of Pain* 7, no. 1 (2006): 43–48. https://doi.org/10.1016/j.jpain.2005.08.001.

Church of Jesus Christ of Latter-day Saints. "Dress and Appearance." Accessed July 2, 2020. https://www.churchofjesuschrist

.org/study/manual/for-the-strength-of-youth/dress-and-appearance
?lang=eng.

Church, Russell M., Vincent LoLordo, J. Bruce Overmier, Richard L.
Solomon, and Lucille H. Turner. "Cardiac Responses to Shock in
Curarized Dogs: Effects of Shock Intensity and Duration, Warning
Signal, and Prior Experience with Shock." *Journal of Comparative
and Physiological Psychology* 62, no. 1 (1966): 1–7. https://doi.org/
10.1037/h0023476.

Cologne, John B., and Dale L. Preston. "Longevity of Atomic-Bomb
Survivors." *Lancet* 356, no. 9226 (2000): 303–7. https://doi.org
/10.1016/S0140-6736(00)02506-X.

Coolen, P., S. Best, A. Lima, J. Sabel, and L. Paulozzi. "Overdose
Deaths Involving Prescription Opioids among Medicaid Enrollees—
Washington, 2004–2007." *Morbidity and Mortality Weekly Report* 58,
no. 42 (2009): 1171–75.

Courtwright, David T. "Addiction to Opium and Morphine."
In *Dark Paradise: A History of Opiate Addiction in America*, 35–60.
Cambridge, MA: Harvard University Press, 2009. https://doi.org
/10.2307/j.ctvk12rb0.7.

Courtwright, David T. *The Age of Addiction: How Bad Habits Became
Big Business*. Cambridge, MA: Belknap Press, 2019. https://doi.org/
10.4159/9780674239241.

Crump, Casey, Kristina Sundquist, Jan Sundquist, and Marilyn A.
Winkleby. "Neighborhood Deprivation and Psychiatric Medication
Prescription: A Swedish National Multilevel Study." *Annals of
Epidemiology* 21, no. 4 (2011): 231–37. https://doi.org/10.1016
/j.annepidem.2011.01.005.

Cui, Changhai, Antonio Noronha, Kenneth R. Warren, George F.
Koob, Rajita Sinha, Mahesh Thakkar, John Matochik, et al. "Brain
Pathways to Recovery from Alcohol Dependence." *Alcohol* 49, no. 5
(2015): 435–52. https://doi.org/10.1016/j.alcohol.2015.04.006.

Cypser, James R., Pat Tedesco, and Thomas E. Johnson. "Hormesis
and Aging in *Caenorhabditis Elegans*." *Experimental Gerontology* 41,
no. 10 (2006): 935–39. https://doi.org/10.1016/j.exger.2006.09.004.

Douthat, Ross. *Bad Religion: How We Became a Nation of Heretics*. New York: Free Press, 2013.

Dunnington, Kent. *Addiction and Virtue: Beyond the Models of Disease and Choice*. Downers Grove, IL: InterVarsity Press Academic, 2011.

Edlund, Mark J., Katherine M. Harris, Harold G. Koenig, Xiaotong Han, Greer Sullivan, Rhonda Mattox, and Lingqi Tang. "Religiosity and Decreased Risk of Substance Use Disorders: Is the Effect Mediated by Social Support or Mental Health Status?" *Social Psychiatry and Psychiatric Epidemiology* 45 (2010): 827–36. https://doi.org/10.1007/s00127-009-0124-3.

Eikelboom, Roelof, and Randelle Hewitt. "Intermittent Access to a Sucrose Solution for Rats Causes Long-Term Increases in Consumption." *Physiology and Behavior* 165 (2016): 77–85. https://doi.org/10.1016/j.physbeh.2016.07.002.

El-Mallakh, Rif S., Yonglin Gao, and R. Jeannie Roberts. "Tardive Dysphoria: The Role of Long Term Antidepressant Use in-Inducing Chronic Depression." *Medical Hypotheses* 76, no. 6 (2011): 769–73. https://doi.org/10.1016/j.mehy.2011.01.020.

Epstein, Mark. *Going on Being: Life at the Crossroads of Buddhism and Psychotherapy*. Boston: Wisdom Publications, 2009.

Fava, Giovanni A., and Fiammetta Cosci. "Understanding and Managing Withdrawal Syndromes after Discontinuation of Antidepressant Drugs." *Journal of Clinical Psychiatry* 80, no. 6 (2019). https://doi.org/10.4088/JCP.19com12794.

Fiorino, Dennis F., Ariane Coury, and Anthony G. Phillips. "Dynamic Changes in Nucleus Accumbens Dopamine Efflux during the Coolidge Effect in Male Rats." *Journal of Neuroscience* 17, no. 12 (1997): 4849–55. https://doi.org/10.1523/jneurosci.17-12-04849.1997.

Fisher, J. P., D. T. Hassan, and N. O'Connor. "Case Report on Pain." *British Medical Journal* 310, no. 6971 (1995): 70. https://www.ncbi.nlm.nih.gov/pmc/articles/PMC2548478/pdf/bmj00574-0074.pdf.

Fogel, Robert William. *The Fourth Great Awakening and the Future of Egalitarianism*. Chicago: University of Chicago Press, 2000.

Francis, David R. "Why High Earners Work Longer Hours." National Bureau of Economic Research digest, 2020. http://www .nber.org/digest/jul06/w11895.html.

Frank, Joseph W., Travis I. Lovejoy, William C. Becker, Benjamin J. Morasco, Christopher J. Koenig, Lilian Hoffecker, Hannah R. Dischinger, et al. "Patient Outcomes in Dose Reduction or Discontinuation of Long-Term Opioid Therapy: A Systematic Review." *Annals of Internal Medicine* 167, no. 3 (2017): 181–91. https://doi.org/ 10.7326/M17-0598.

Franken, Ingmar H. A., Corien Zijlstra, and Peter Muris. "Are Nonpharmacological Induced Rewards Related to Anhedonia? A Study among Skydivers." *Progress in Neuro-Psychopharmacology and Biological Psychiatry* 30, no. 2 (2006): 297–300. https://doi.org /10.1016/j.pnpbp.2005.10.011.

Gasparro, Annie, and Jessie Newman. "The New Science of Taste: 1,000 Banana Flavors." *Wall Street Journal*, October 31, 2014.

Ghertner, Robin, and Lincoln Groves. "The Opioid Crisis and Economic Opportunity: Geographic and Economic Trends." ASPE Research Brief from the U.S. Department of Health and Human Services, 2018. https://aspe.hhs.gov/system/files/pdf/259261 /ASPEEconomicOpportunityOpioidCrisis.pdf.

Grant, Bridget F., S. Patricia Chou, Tulshi D. Saha, Roger P. Pickering, Bradley T. Kerridge, W. June Ruan, Boji Huang, et al. "Prevalence of 12-Month Alcohol Use, High-Risk Drinking, and DSM-IV Alcohol Use Disorder in the United States, 2001–2002 to 2012–2013: Results from the National Epidemiologic Survey on Alcohol and Related Conditions." *JAMA Psychiatry* 74, no. 9 (September 1, 2017): 911–23. https://doi.org/10.1001/ jamapsychiatry.2017.2161.

Hall, Wayne. "What Are the Policy Lessons of National Alcohol Prohibition in the United States, 1920–1933?" *Addiction* 105, no. 7 (2010): 1164–73. https://doi.org/10.1111/j.1360-0443.2010.02926.x.

Hatcher, Alexandrea E., Sonia Mendoza, and Helena Hansen. "At the Expense of a Life: Race, Class, and the Meaning of Buprenorphine in Pharmaceuticalized 'Care.'" *Substance Use and Misuse* 53, no. 2 (2018): 301–10. https://doi.org/10.1080/10826084.2017.1385633.

Helliwell, John F., Haifang Huang, and Shun Wang. "Chapter 2: Changing World Happiness." *World Happiness Report 2019*, March 20, 2019.

Hippocrates. *Aphorisms*. Accessed July 8, 2020. http://classics.mit .edu/Hippocrates/aphorisms.1.i.html.

Howie, Lajeana D., Patricia N. Pastor, and Susan L. Lukacs. "Use of Medication Prescribed for Emotional or Behavioral Difficulties Among Children Aged 6–17 Years in the United States, 2011–2012." *Health Care in the United States: Developments and Considerations* 5, no. 148 (2015): 25–35.

Hung, Lin W., Sophie Neuner, Jai S. Polepalli, Kevin T. Beier, Matthew Wright, Jessica J. Walsh, Eastman M. Lewis, et al. "Gating of Social Reward by Oxytocin in the Ventral Tegmental Area." *Science* 357, no. 6358 (2017): 1406–11. https://doi.org/10.1126/science.aan4994.

Huxley, Aldous. *Brave New World Revisited*. New York: HarperCollins, 2004.

Iannaccone, Laurence R. "Sacrifice and Stigma: Reducing Free-Riding in Cults, Communes, and Other Collectives." *Journal of Political Economy* 100, no. 2 (1992): 271–91.

Iannaccone, Laurence R. "Why Strict Churches Are Strong." *American Journal of Sociology* 99, no. 5 (1994): 1180–1211. https:// doi.org/10.2307/2781147.

Iannelli, Eric J. "Species of Madness." *Times Literary Supplement*, September 22, 2017.

Jonas, Bruce S., Qiuping Gu, and Juan R. Albertorio-Diaz. "Psychotropic Medication Use among Adolescents: United States, 2005–2010." *NCHS Data Brief*, no. 135 (December 2013): 1–8.

Jorm, Anthony F., Scott B. Patten, Traolach S. Brugha, and Ramin Mojtabai. "Has Increased Provision of Treatment Reduced the Prevalence of Common Mental Disorders? Review of the Evidence from Four Countries." *World Psychiatry* 16, no. 1 (2017): 90–99. https://doi.org/10.1002/wps.20388.

Kant, Immanuel. "Groundwork of the Metaphysic of Morals (1785)," *Cambridge Texts in the History of Philosophy.* Cambridge: Cambridge University Press, 1998.

Katcher, Aaron H., Richard L. Solomon, Lucille H. Turner, and Vincent Lolordo. "Heart Rate and Blood Pressure Responses to Signaled and Unsignaled Shocks: Effects of Cardiac Sympathectomy." *Journal of Comparative and Physiological Psychology* 68, no. 2 (1969): 163–74.

Kidd, Celeste, Holly Palmeri, and Richard N. Aslin. "Rational Snacking: Young Children's Decision-Making on the Marshmallow Task Is Moderated by Beliefs about Environmental Reliability." *Cognition* 126, no. 1 (2013): 109–14. https://doi.org/10.1016 /j.cognition.2012.08.004.

Knibbs, Kate. "All the Gear an Ultramarathoner Legend Brings with Him on the Trail." Gizmodo, October 29, 2015. https://gizmodo.com /all-the-gear-an-ultramarathon-legend-brings-with-him-on -1736088954.

Kohrman, Matthew, Quan Gan, Liu Wennan, and Robert N. Proctor, eds. *Poisonous Pandas: Chinese Cigarette Manufacturing in Critical Historical Perspectives.* Stanford, CA: Stanford University Press, 2018.

Kolb, Brian, Grazyna Gorny, Yilin Li, Anne-Noël Samaha, and Terry E. Robinson. "Amphetamine or Cocaine Limits the Ability of Later Experience to Promote Structural Plasticity in the Neocortex and Nucleus Accumbens." *Proceedings of the National Academy of Sciences of the United States of America* 100, no. 18 (2003): 10523–28. https://doi.org/10.1073/pnas.1834271100.

Koob, George F. "Hedonic Homeostatic Dysregulation as a Driver of Drug-Seeking Behavior." *Drug Discovery Today: Disease Models* 5, no. 4 (2008): 207–15. https://doi.org/10.1016/j.ddmod.2009.04.002.

Kramer, Peter D. *Listening to Prozac.* New York: Viking Press, 1993.

Kreher, Jeffrey B., and Jennifer B. Schwartz. "Overtraining Syndrome: A Practical Guide." *Sports Health* 4, no. 2 (2012). https://doi.org/10.1177/1941738111434406.

Leknes, Siri, and Irene Tracey. "A Common Neurobiology for Pain and Pleasure." *Nature Reviews Neuroscience* 9, no. 4 (2008): 314–20. https://doi.org/10.1038/nrn2333.

Lembke, Anna. *Drug Dealer, MD: How Doctors Were Duped, Patients Got Hooked, and Why It's So Hard to Stop.* 1st ed. Baltimore: Johns Hopkins University Press, 2016.

Lembke, Anna. "Time to Abandon the Self-Medication Hypothesis in Patients with Psychiatric Disorders." *American Journal of Drug and Alcohol Abuse* 38, no. 6 (2012): 524–29. https://doi.org/10.3109/00952990.2012.694532.

Lembke, Anna, and Amer Raheemullah. "Addiction and Exercise." In *Lifestyle Psychiatry: Using Exercise, Diet and Mindfulness to Manage Psychiatric Disorders*, edited by Doug Noordsy. Washington, DC: American Psychiatric Publishing, 2019.

Lembke, Anna, and Niushen Zhang. "A Qualitative Study of Treatment-Seeking Heroin Users in Contemporary China." *Addiction Science & Clinical Practice* 10, no. 23 (2015). https://doi.org/10.1186/s13722-015-0044-3.

Levin, Edmund C. "The Challenges of Treating Developmental Trauma Disorder in a Residential Agency for Youth." *Journal of the American Academy of Psychoanalysis and Dynamic Psychiatry* 37, no. 3 (2009): 519–38. https://doi.org/10.1521/jaap.2009.37.3.519.

Linnet, J., E. Peterson, D. J. Doudet, A. Gjedde, and A. Møller. "Dopamine Release in Ventral Striatum of Pathological Gamblers Losing Money." *Acta Psychiatrica Scandinavica* 122, no. 4 (2010): 326–33. https://doi.org/10.1111/j.1600-0447.2010.01591.x.

Liu, Qingqing, Hairong He, Jin Yang, Xiaojie Feng, Fanfan Zhao, and Jun Lyu. "Changes in the Global Burden of Depression from 1990 to 2017: Findings from the Global Burden of Disease Study." *Journal of*

Psychiatric Research 126 (June 2020): 134–40. https://doi.org/10.1016/j.jpsychires.2019.08.002.

Liu, Xiang. "Inhibiting Pain with Pain—A Basic Neuromechanism of Acupuncture Analgesia." *Chinese Science Bulletin* 46, no. 17 (2001): 1485–94. https://doi.org/10.1007/BF03187038.

Low, Yinghui, Collin F. Clarke, and Billy K. Huh. "Opioid-Induced Hyperalgesia: A Review of Epidemiology, Mechanisms and Management." *Singapore Medical Journal* 53, no. 5 (2012): 357–60.

MacCoun, Robert. "Drugs and the Law: A Psychological Analysis of Drug Prohibition." *Psychological Bulletin* 113 (June 1, 1993): 497–512. https://doi.org/10.1037//0033-2909.113.3.497.

Maréchal, Michel André, Alain Cohn, Giuseppe Ugazio, and Christian C. Ruff. "Increasing Honesty in Humans with Noninvasive Brain Stimulation." *Proceedings of the National Academy of Sciences of the United States of America* 114, no. 17 (2017): 4360–64. https://doi.org/10.1073/pnas.1614912114.

Mattson, Mark P. "Energy Intake and Exercise as Determinants of Brain Health and Vulnerability to Injury and Disease." *Cell Metabolism* 16, no. 6 (2012): 706–22. https://doi.org/10.1016/j.cmet.2012.08.012.

Mattson, Mark P., and Ruiqian Wan. "Beneficial Effects of Intermittent Fasting and Caloric Restriction on the Cardiovascular and Cerebrovascular Systems." *Journal of Nutritional Biochemistry* 16, no. 3 (2005): 129–37. https://doi.org/10.1016/j.jnutbio.2004.12.007.

McClure, Samuel M., David I. Laibson, George Loewenstein, and Jonathan D. Cohen. "Separate Neural Systems Value Immediate and Delayed Monetary Rewards." *Science* 306, no. 5695 (2004): 503–7. https://doi.org/10.1126/science.1100907.

Meijer, Johanna H., and Yuri Robbers. "Wheel Running in the Wild." *Proceedings of the Royal Society B: Biological Sciences*, July 7, 2014. https://doi.org/10.1098/rspb.2014.0210.

Meldrum, M. L. "A Capsule History of Pain Management." *JAMA* 290, no. 18 (2003): 2470–75. https://doi.org/10.1001/jama.290.18.2470.

Mendis, Shanthi, Tim Armstrong, Douglas Bettcher, Francesco Branca, Jeremy Lauer, Cecile Mace, Shanthi Mendis, et al. *Global Status Report on Noncommunicable Diseases 2014*. World Health Organization, 2014. https://apps.who.int/iris/bitstream /handle/10665/148114/9789241564854_eng.pdf.

Minois, Nadège. "The Hormetic Effects of Hypergravity on Longevity and Aging." *Dose-Response* 4, no. 2 (2006). https://doi.org/ 10.2203/dose-response.05-008.minois.

Montagu, Kathleen A. "Catechol Compounds in Rat Tissues and in Brains of Different Animals." *Nature* 180 (1957): 244–45. https://doi.org/10.1038/180244a0.

National Potato Council. *Potato Statistical Yearbook 2016*. Accessed April 18, 2020. https://www.nationalpotatocouncil.org/files/7014 /6919/7938/NPCyearbook2016_-_FINAL.pdf.

Ng, Marie, Tom Fleming, Margaret Robinson, Blake Thomson, Nicholas Graetz, Christopher Margono, Erin C. Mullany, et al. "Global, Regional, and National Prevalence of Overweight and Obesity in Children and Adults during 1980–2013: A Systematic Analysis for the Global Burden of Disease Study 2013." *Lancet* 384, no. 9945 (August 2014): 766–81. https://doi.org/10.1016/ S0140-6736(14)60460-8.

Ng, S. W., and B. M. Popkin. "Time Use and Physical Activity: A Shift Away from Movement across the Globe," *Obesity Reviews* 13, no. 8 (August 2012): 659–80. https://doi.org/10.1111/j.1467-789X.2011 .00982.x.

O'Dell, Laura E., Scott A. Chen, Ron T. Smith, Sheila E. Specio, Robert L. Balster, Neil E. Paterson, Athina Markou, et al. "Extended Access to Nicotine Self-Administration Leads to Dependence: Circadian Measures, Withdrawal Measures, and Extinction Behavior in Rats." *Journal of Pharmacology and Experimental Therapeutics* 320, no. 1 (2007): 180–93. https://doi.org/10.1124/ jpet.106.105270.

OECD. "OECD Health Statistics 2020," July 2020. http://www.oecd .org/els/health-systems/health-data.htm.

OECD. "Special Focus: Measuring Leisure in OECD Countries." In *Society at a Glance 2009: OECD Social Indicators*. Paris: OECD Publishing, 2009. https://doi.org/10.1787/soc_glance-2008-en.

Ohe, Christina G. von der, Corinna Darian-Smith, Craig C. Garner, and H. Craig Heller. "Ubiquitous and Temperature-Dependent Neural Plasticity in Hibernators." *Journal of Neuroscience* 26, no. 41 (2006): 10590–98. https://doi.org/10.1523/JNEUROSCI.2874-06.2006.

Omura, Daniel T., Damon A. Clark, Aravinthan D. T. Samuel, and H. Robert Horvitz. "Dopamine Signaling Is Essential for Precise Rates of Locomotion by C. Elegans." *PLOS ONE* 7, no. 6 (2012). https://doi.org/10.1371/journal.pone.0038649.

Östlund, Magdalena Plecka, Olof Backman, Richard Marsk, Dag Stockeld, Jesper Lagergren, Finn Rasmussen, and Erik Näslund. "Increased Admission for Alcohol Dependence after Gastric Bypass Surgery Compared with Restrictive Bariatric Surgery." *JAMA Surgery* 148, no. 4 (2013): 374–77. https://doi.org/10.1001/jamasurg.2013.700.

Pascoli, Vincent, Marc Turiault, and Christian Lüscher. "Reversal of Cocaine-Evoked Synaptic Potentiation Resets Drug-Induced Adaptive Behaviour." *Nature* 481 (2012): 71–75. https://doi.org/10.1038/nature10709.

Pedersen, B. K., and B. Saltin. "Exercise as Medicine—Evidence for Prescribing Exercise as Therapy in 26 Different Chronic Diseases." *Scandinavian Journal of Medicine and Science in Sports* 25, no. S3 (2015): 1–72.

Petry, Nancy M., Warren K. Bickel, and Martha Arnett. "Shortened Time Horizons and Insensitivity to Future Consequences in Heroin Addicts." *Addiction* 93, no. 5 (1998): 729–38. https://doi.org/10.1046/j.1360-0443.1998.9357298.x.

Piper, Brian J., Christy L. Ogden, Olapeju M. Simoyan, Daniel Y. Chung, James F. Caggiano, Stephanie D. Nichols, and Kenneth L. McCall. "Trends in Use of Prescription Stimulants in the United States and Territories, 2006 to 2016." *PLOS ONE* 13, no. 11 (2018). https://doi.org/10.1371/journal.pone.0206100.

Postman, Neil. *Amusing Ourselves to Death: Public Discourse in the Age of Show Business.* New York: Penguin Books, 1986.

Pratt, Laura A., Debra J. Brody, and Quiping Gu. "Antidepressant Use in Persons Aged 12 and Over: United States, 2005–2008." *NCHS Data Brief No. 76,* October 2011. https://www.cdc.gov/nchs/products /databriefs/db76.htm.

"Qur'an: Verse 24:31." Accessed July 2, 2020. http://corpus.quran .com/translation.jsp?chapter=24&verse=31.

Ramos, Dandara, Tânia Victor, Maria Lucia Seidl-de-Moura, and Martin Daly. "Future Discounting by Slum-Dwelling Youth versus University Students in Rio de Janeiro." *Journal of Research on Adolescence* 23, no. 1 (2013): 95–102. https://doi.org/10.1111/j.1532-7795 .2012.00796.x.

Rieff, Philip. *The Triumph of the Therapeutic: Uses of Faith after Freud.* New York: Harper and Row, 1966.

Ritchie, Hannah, and Max Roser. "Drug Use." Our World in Data. Retrieved 2019. https://ourworldindata.org/drug-use.

Robinson, Terry E., and Bryan Kolb. "Structural Plasticity Associated with Exposure to Drugs of Abuse." *Neuropharmacology* 47, Suppl. 1 (2004): 33–46. https://doi.org/10.1016/j.neuropharm.2004.06.025.

Rogers, J. L., S. De Santis, and R. E. See. "Extended Methamphet-amine Self-Administration Enhances Reinstatement of Drug Seeking and Impairs Novel Object Recognition in Rats." *Psychopharmacology* 199, no. 4 (2008): 615–24. https://doi.org/10.1007/s00213-008-1187-7.

Ruscio, Ayelet Meron, Lauren S. Hallion, Carmen C. W. Lim, Sergio Aguilar-Gaxiola, Ali Al-Hamzawi, Jordi Alonso, Laura Helena Andrade, et al. "Cross-Sectional Comparison of the Epidemiology of *DSM-5* Generalized Anxiety Disorder across the Globe." *JAMA Psychiatry* 74, no. 5 (2017): 465–75. https://doi.org/10.1001 /jamapsychiatry.2017.0056.

Saal, Daniel, Yan Dong, Antonello Bonci, and Robert C. Malenka. "Drugs of Abuse and Stress Trigger a Common Synaptic Adaptation in Dopamine Neurons." *Neuron* 37, no. 4 (2003): 577–82. https://doi .org/10.1016/S0896-6273(03)00021-7.

Satel, Sally, and Scott O. Lilienfeld. "Addiction and the Brain-Disease Fallacy." *Frontiers in Psychiatry* 4 (March 2014): 1–11. https://doi.org/10.3389/fpsyt.2013.00141.

Schelling, Thomas. "Self-Command in Practice, in Policy, and in a Theory of Rational Choice." *American Economic Review* 74, no. 2 (1984): 1–11. https://econpapers.repec.org/article/aeaaecrev/v_3a74 _3ay_3a1984_3ai_3a2_3ap_3a1-11.htm. https://doi.org/10.2307/1816322.

Schwarz, Alan. "Thousands of Toddlers Are Medicated for A.D.H.D., Report Finds, Raising Worries." *New York Times*, May 16, 2014.

Shanmugam, Victoria K., Kara S. Couch, Sean McNish, and Richard L. Amdur. "Relationship between Opioid Treatment and Rate of Healing in Chronic Wounds." *Wound Repair and Regeneration* 25, no. 1 (2017): 120–30. https://doi.org/10.1111/wrr.12496.

Sharp, Mark J., and Thomas A. Melnik. "Poisoning Deaths Involving Opioid Analgesics—New York State, 2003–2012." *Morbidity and Mortality Weekly Report* 64, no. 14 (2015): 377–80.

Shahbandeh, M. "Gluten-Free Food Market Value in the United States from 2014 to 2025." Statista, November 20, 2019. Accessed July 2, 2020. https://www.statista.com/statistics/884086/us-gluten -free-food-market-value/.

Sherwin, C. M. "Voluntary Wheel Running: A Review and Novel Interpretation." *Animal Behaviour* 56, no. 1 (1998): 11–27. https://doi .org/10.1006/anbe.1998.0836.

Shoda, Yuichi, Walter Mischel, and Philip K. Peake. "Predicting Adolescent Cognitive and Self-Regulatory Competencies from Preschool Delay of Gratification: Identifying Diagnostic Conditions." *Developmental Psychology* 26, no. 6 (1990): 978–86. https://doi.org/10.1037/0012-1649.26.6.978.

Sinclair, J. D. "Evidence about the Use of Naltrexone and for Different Ways of Using It in the Treatment of Alcoholism." *Alcohol and Alcoholism* 36, no. 1 (2001): 2–10. https://doi.org/10.1093/alcalc/36.1.2.

Singh, Amit, and Sujita Kumar Kar. "How Electroconvulsive Therapy Works?: Understanding the Neurobiological Mechanisms." *Clinical Psychopharmacology and Neuroscience* 15, no. 3 (2017): 210–21. https://doi.org/10.9758/cpn.2017.15.3.210.

Sobell, L. C., J. A. Cunningham, and M. B. Sobell. "Recovery from Alcohol Problems with and without Treatment: Prevalence in Two Population Surveys." *American Journal of Public Health* 86, no. 7 (1996): 966–72.

Sobell, Mark B., and Linda C. Sobell. "Controlled Drinking after 25 Years: How Important Was the Great Debate?" *Addiction* 90, no. 9 (1995): 1149–53.

Solomon, Richard L., and John D. Corbit. "An Opponent-Process Theory of Motivation." *American Economic Review* 68, no. 6 (1978): 12–24.

Spoelder, Marcia, Peter Hesseling, Annemarie M. Baars, José G. Lozeman-van't Klooster, Marthe D. Rotte, Louk J. M. J. Vanderschuren, and Heidi M. B. Lesscher. "Individual Variation in Alcohol Intake Predicts Reinforcement, Motivation, and Compulsive Alcohol Use in Rats." *Alcoholism: Clinical and Experimental Research* 39, no. 12 (2015): 2427–37. https://doi.org/10.1111/acer.12891.

Sprenger, Christian, Ulrike Bingel, and Christian Büchel. "Treating Pain with Pain: Supraspinal Mechanisms of Endogenous Analgesia Elicited by Heterotopic Noxious Conditioning Stimulation." *Pain* 152, no. 2 (2011): 428–39. https://doi.org/10.1016/j.pain.2010.11.018.

Šrámek, P., M. Šimečková, L. Janský, J. Šavlíková, and S. Vybíral. "Human Physiological Responses to Immersion into Water of Different Temperatures." *European Journal of Applied Physiology* 81 (2000): 436–42. https://doi.org/10.1007/s004210050065.

Strang, John, Thomas Babor, Jonathan Caulkins, Benedikt Fischer, David Foxcroft, and Keith Humphreys. "Drug Policy and the Public Good: Evidence for Effective Interventions." *Lancet* 379 (2012): 71–83.

Substance Abuse and Mental Health Services Administration, U.S. Department of Health and Human Services. *Behavioral Health,*

United States, 2012. HHS Publication No. (SMA) 13-4797, 2013. http://www.samhsa.gov/data/sites/default/files/2012-BHUS.pdf.

Sutou, Shizuyo. "Low-Dose Radiation from A-Bombs Elongated Lifespan and Reduced Cancer Mortality Relative to Un-Irradiated Individuals." *Genes and Environment* 40, no. 26 (2018). https://doi .org/10.1186/s41021-018-0114-3.

Sydenham, Thomas. "A Treatise of the Gout and Dropsy." In *The Works of Thomas Sydenham, M.D., on Acute and Chronic Diseases,* 254. London, 1783. https://books.google.com/books?id=iSxsAAAA -MAAJ&printsec=frontcover&source=gbs_ge_summary_r&cad= 0#v=onepage&q&f=false 2.

Synnott, Mark. "How Alex Honnold Made the Ultimate Climb without a Rope." *National Geographic* online. Accessed July 8, 2020. https://www.nationalgeographic.com/magazine/2019/02/alex -honnold-made-ultimate-climb-el-capitan-without-rope.

Synnott, Mark. *The Impossible Climb: Alex Honnold, El Capitan, and the Climbing Life.* New York: Dutton, 2018.

Taussig, Helen B. "'Death' from Lightning and the Possibility of Living Again." *American Scientist* 57, no. 3 (1969): 306–16.

Tomek, Seven E., Gabriela M. Stegmann, and M. Foster Olive. "Effects of Heroin on Rat Prosocial Behavior." *Addiction Biology* 24, no. 4 (2019): 676–84. https://doi.org/10.1111/adb.12633.

Twelve Steps and Twelve Traditions. New York: Alcoholics Anonymous World Services, n.d.

Vasconcellos, Silvio José Lemos, Matheus Rizzatti, Thamires Pereira Barbosa, Bruna Sangoi Schmitz, Vanessa Cristina Nascimento Coelho, and Andrea Machado. "Understanding Lies Based on Evolutionary Psychology: A Critical Review." *Trends in Psychology* 27, no. 1 (2019): 141–53. https://doi.org/10.9788/TP2019.1-11.

Vengeliene, Valentina, Ainhoa Bilbao, and Rainer Spanagel. "The Alcohol Deprivation Effect Model for Studying Relapse Behavior: A Comparison between Rats and Mice." *Alcohol* 48, no. 3 (2014): 313–20. https://doi.org/10.1016/j.alcohol.2014.03.002.

Volkow, N. D., J. S. Fowler, and G. J. Wang. "Role of Dopamine in Drug Reinforcement and Addiction in Humans: Results from Imaging Studies." *Behavioural Pharmacology* 13, no. 5 (2002): 355–66. https://doi.org/10.1097/00008877-200209000-00008.

Volkow, N. D., J. S. Fowler, G-J. Wang, and J. M. Swanson. "Dopamine in Drug Abuse and Addiction: Results from Imaging Studies and Treatment Implications." *Molecular Psychiatry* 9, no. 6 (June 2004): 557–69. https://doi.org/10.1038/sj.mp.4001507.

Watson, Gretchen LeFever, Andrea Powell Arcona, and David O. Antonuccio. "The ADHD Drug Abuse Crisis on American College Campuses." *Ethical Human Psychology and Psychiatry* 17, no. 1 (2015). https://doi.org/10.1891/1559-4343.17.1.5.

Weisman, Aly, and Kristen Griffin. "Jimmy Kimmel Lost a Ton of Weight on This Radical Diet." *Business Insider,* January 9, 2016.

Wells, K. B., R. Sturm, C. D. Sherbourne, and L. S. Meredith. *Caring for Depression.* Cambridge, MA: Harvard University Press, 1996.

Whitaker, Robert. *Anatomy of an Epidemic: Magic Bullets, Psychiatric Drugs, and the Astonishing Rise of Mental Illness in America.* New York: Crown, 2010.

Winnicott, Donald W. "Ego Distortion in Terms of True and False Self." In *The Maturational Process and the Facilitating Environment: Studies in the Theory of Emotional Development,* 140–57. New York: International Universities Press, 1960.

Wu, Tim. "The Tyranny of Convenience." *New York Times,* February 6, 2018.

Younger, Jarred, Noorulain Noor, Rebecca McCue, and Sean Mackey. "Low-Dose Naltrexone for the Treatment of Fibromyalgia: Findings of a Small, Randomized, Double-Blind, Placebo-Controlled, Counter-balanced, Crossover Trial Assessing Daily Pain Levels." *Arthritis and Rheumatism* 65, no. 2 (2013): 529–38. https://doi.org/10.1002/art.37734.

Zhou, Qun Yong, and Richard D. Palmiter. "Dopamine-Deficient Mice Are Severely Hypoactive, Adipsic, and Aphagic." *Cell* 83, no. 7 (1995): 1197–1209. https://doi.org/10.1016/0092-8674(95)90145-0.

Acknowledgments

I would like to thank my patients who shared their experiences and reflections with me in the process of writing this book. Their willingness to give of themselves not just to me but also to unseen, unknown readers, is an act of courage and generosity. This is our book.

I would also like to thank the people who are not my patients who agreed to be interviewed for this book. Their insights on addiction and recovery have added immeasurably to my own.

I'm fortunate to be surrounded by many thoughtful and creative people whose ideas have made their way into this book through our conversations. It would be impossible to list them all, but I want to extend a special thank-you to Kent Dunnington, Keith Humphreys, E. J. Iannelli, Rob Malenka, Matthew Prekupec, John Ruark, and Daniel Saal.

Thanks also to Robin Coleman for getting me writing again, Bonnie Solow for believing in the project, Deb McCarroll for painting the pictures, and Stephen Morrow and Hannah Feeney for bringing it to fruition.

Finally, nothing would be possible without the support of my beloved husband, Andrew.

Index

Note: Page numbers in *italics* refer to illustrations.